SpringerBriefs in Applied Sciences and Technology

Computational Intelligence

Series editor

Janusz Kacprzyk, Warsaw, Poland

For further volumes:
http://www.springer.com/series/10618

About this Series

The series "Studies in Computational Intelligence" (SCI) publishes new developments and advances in the various areas of computational intelligence—quickly and with a high quality. The intent is to cover the theory, applications, and design methods of computational intelligence, as embedded in the fields of engineering, computer science, physics and life sciences, as well as the methodologies behind them. The series contains monographs, lecture notes and edited volumes in computational intelligence spanning the areas of neural networks, connectionist systems, genetic algorithms, evolutionary computation, artificial intelligence, cellular automata, self-organizing systems, soft computing, fuzzy systems, and hybrid intelligent systems. Of particular value to both the contributors and the readership are the short publication timeframe and the world-wide distribution, which enable both wide and rapid dissemination of research output.

Oliver Kramer

A Brief Introduction to Continuous Evolutionary Optimization

 Springer

Oliver Kramer
Department für Informatik
Carl von Ossietzky University of Oldenburg
Oldenburg
Germany

ISSN 2191-530X   ISSN 2191-5318   (electronic)
ISBN 978-3-319-03421-8   ISBN 978-3-319-03422-5   (eBook)
DOI 10.1007/978-3-319-03422-5
Springer Cham Heidelberg New York Dordrecht London

Library of Congress Control Number: 2013954569

Printed on acid-free paper

Springer is part of Springer Science+Business Media (www.springer.com)

Acknowledgments

I thank the German Research Foundation (Deutsche Forschungsgemeinschaft, DFG) and the German Academic Exchange Service (Deutscher Akademischer Austauschdienst, DAAD) for the support in the last three years. Further, I thank Günter Rudolph for the support and the discussions about evolutionary computation, and the International Computer Science Institute (ICSI) in Berkeley for their excellent support during my stay in California in 2010, in particular Richard Karp and Nelson Morgan. I thank the Bauhaus-Universität Weimar for their support in 2011. Last, I want to thank all graduate students, postdocs, professors, and other scientists I have been collaborating with since the beginning of my research activities. I had numerous interesting discussions and research collaborations with Jörg Bremer, Fabian Gieseke, Justin Heinermann, Christian Hinrichs, Sascha Hunold, Jörg Lässig, Jendrik Poloczek, Benjamin Satzger, Dirk Sudholt, and Nils André Treiber.

Cancun, June 2013   Oliver Kramer

Contents

Part II   Advanced Optimization

Part III   Learning

Abstract

Practical optimization problems are often hard to solve, in particular when they are black boxes and no further information about the problem is available except via function evaluations. This work introduces a collection of heuristics and algorithms for black box optimization with evolutionary algorithms in continuous solution spaces. The book gives an introduction to evolution strategies and parameter control. Heuristic extensions are presented that allow optimization in constrained, multimodal, and multiobjective solution spaces. An adaptive penalty function is introduced for constrained optimization. Meta-models reduce the number of fitness and constraint function calls in expensive optimization problems. The hybridization of evolution strategies with local search allows fast optimization in solution spaces with many local optima. A selection operator based on reference lines in objective space is introduced to optimize multiple conflictive objectives. Evolutionary search is employed for learning kernel parameters of the Nadaraya-Watson estimator, and a swarm-based iterative approach is presented for optimizing latent points in dimensionality reduction problems. Experiments on typical benchmark problems as well as numerous figures and diagrams illustrate the behavior of the introduced concepts and methods.

Part I
Foundations

Chapter 1
Introduction

Many optimization problems that have to be solved in practice are black box problems. Often, not much is known about an optimization problem except the information one can get via function evaluations. Neither derivatives nor constraints are known. In the worst case, nothing is even known about the characteristics of the fitness function, e.g., whether it is uni- or multimodal. This scenario affords the application of specialized optimization strategies often called direct search methods. Evolutionary algorithms that mimic the biological notion of evolution and employ stochastic components to search in the solution space have grown to strong optimization methods. Evolutionary methods that are able to efficiently search in large optimization scenarios and learn from observed patterns in data mining scenarios have found broad acceptance in many disciplines, e.g., civil and electrical engineering. The methods have been influenced from various disciplines: robotics, statistics, computer science, engineering, and the cognitive sciences. This might be the reason for the large variety of techniques that have been developed in the last decades. The employment of computer simulations has become outstandingly successful in engineering within the last years. This development includes the application of optimization and learning techniques in the design and prototype process. Simulations allow the study of prototype characteristics before the product has actually been manufactured. Such a process allows an entirely computed-based optimization of the whole prototype or of its parts and can result in significant speedups and savings of material and money.

Learning and optimization are strongly related to each other. In optimization, one seeks for optimal parameters of a function or system w.r.t. a defined objective. In machine learning, one seeks for an optimal functional model that allows to describe relations between observations. Pattern recognition and machine learning problems also involve solving optimization problems. Many different optimization approaches are employed, from heuristics with stochastic components to exact convex methods.

The goal of this book is to give a brief introduction to latest heuristics in evolutionary optimization for continuous solution spaces. The beginning of the work gives a short introduction to the main problem classes of interest: optimization, super-, and

O. Kramer, *A Brief Introduction to Continuous Evolutionary Optimization*,
SpringerBriefs in Computational Intelligence,
DOI: 10.1007/978-3-319-03422-5_1, © The Author(s) 2014

Fig. 1.1 Survey of problem classes the methods in this work belong to: evolutionary optimization, super-, and unsupervised learning

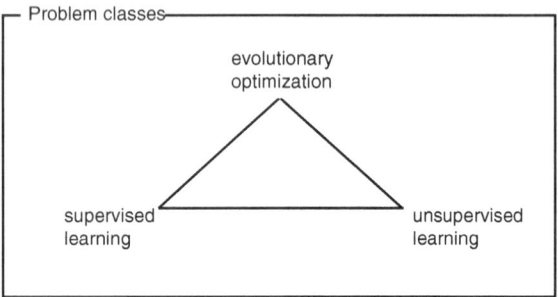

unsupervised learning, see Fig. 1.1. Optimization is the problem of finding optimal parameters for arbitrary models and functions. Supervised learning is about finding functional models that best model observations with given label information. Unsupervised learning is about learning functional models only based on the structure of the data itself, i.e., without label information. The following three paragraphs give a short introduction to the three problem classes.

1.1 Optimization

Optimization is the search for optimal parameters of a system. The parameters are known as design or objective variables. We assume a set S of solutions that we call solution space or search space. A typical example for a solution space is the set $\mathbb{R}$ of continuous values. In most cases, not only one, but many values have to be optimized at the same time resulting in an N-dimensional search problem, or search problem dimensionality, respectively. For continuous solution spaces, this means we search in $\mathbb{R}^N$. A famous optimization problem is the traveling salesperson problem. The salesperson has to find the shortest tour through a set of cities and go back to the city, where he started from. In this scenario, a solution consists of a sequence of cities. A feasible solution must contain all cities. Obviously, the solution space has a different structure than the set of continuous solutions. For such solution spaces, special operators have to be employed. We focus on continuous optimization in this book.

Optimality can only be defined w.r.t. some quality measure. We measure the quality of a solution with the help of a quality function f that we also call fitness function. An optimal solution $\mathbf{x}^*$ has a better fitness $f(\mathbf{x}^*)$ than all other solutions $\mathbf{x}$ in the solution space $\mathbb{R}^N$, i.e., for an optimal solution $\mathbf{x}^* \in \mathbb{R}^N$ it holds $f(\mathbf{x}^*) \leq f(\mathbf{x})$ for all $\mathbf{x} \in \mathbb{R}^N$. This definition holds for single-objective optimization problems and has to be extended for multi-objective problems via the concept of Pareto optimality, see Chap. 6. Without loss of generality, I concentrate on minimization problems. Maximization problems can easily be transformed into minimization problems by inversion of the objective function

$$f_{\min}(\mathbf{x}) = -f_{\max}(\mathbf{x}). \tag{1.1}$$

A solution $\mathbf{x}^*$ with a better fitness $f(\mathbf{x}^*) < f(\mathbf{x})$ than the solutions in its environment $\mathbf{x} \in \mathbb{R}^N$ with $\|\mathbf{x} - \mathbf{x}^*\| < \epsilon$ for an $\epsilon > 0$ is called local optimum. Objective variables can be discrete (i.e., they are defined over a discrete set) or continuous (defined over $\mathbb{R}^N$). This work concentrates on continuous black box optimization problems, where no derivatives or functional expressions are explicitly given. This is a reasonable assumption, as many optimization problems in practice are black boxes.

For unconstrained solution spaces, the conditions for optimal solutions can also be formulated via the first and the second partial derivatives. The necessary condition for a solution $\mathbf{x}^*$ to be a minimum is that the gradient vanishes at $\mathbf{x}^*$, i.e., $\nabla f(\mathbf{x}^*) = \mathbf{0}$. To exclude saddle points, the second derivative at $\mathbf{x}^*$ has to change the sign. The condition can conveniently be formulated with the Hessian matrix $\mathbf{H}$ that comprises the second partial derivatives

$$\mathbf{H}(f) = \nabla^2 f = \left[\frac{\partial^2 f}{\partial x_i \partial x_j} \right]_{i,j=1,\dots,N} \tag{1.2}$$

when $\mathbf{H}$ is positive-definite, i.e., $\mathbf{x}^T \mathbf{H} \mathbf{x}$ is positive for all non-zero column vectors $\mathbf{x} \in \mathbb{R}^N$. In many non-black box cases, an analytic solution via the above conditions can be obtained by solving the equations. This work concentrates on black box scenarios, where $\mathbf{H}$ is not available.

If certain conditions for objective functions and constraints are fulfilled, the optimization problem can efficiently be solved. In case of linearity of objective functions and constraints, i.e., the objective function is of the form

$$f(\mathbf{x}) = \mathbf{c}^T \mathbf{x}, \tag{1.3}$$

with a vector $\mathbf{c} \in \mathbb{R}^N$, decision variable vector $\mathbf{x} \in \mathbb{R}^N$, and constraints of the form

$$g_i(\mathbf{x}) = \mathbf{A}\mathbf{x} + \mathbf{b} \le 0, \quad i = 1, \dots, n_1, \tag{1.4}$$

and

$$h_j(\mathbf{x}) = \mathbf{B}\mathbf{x} + \mathbf{d} = 0, \quad j = 1, \dots, n_2, \tag{1.5}$$

with matrices $\mathbf{A}, \mathbf{B} \in \mathbb{R}^{N \times N}$, the problem is called linear programming problems. It can efficiently be solved with sequential linear programming techniques. If the objective function is quadratic, i.e.,

$$f(\mathbf{x}) = \frac{1}{2}\mathbf{x}^T \mathbf{H} \mathbf{x} + \mathbf{c}^T \mathbf{x}, \tag{1.6}$$

under the above constraints 1.4 and 1.5, sequential quadratic programming techniques can be applied to efficiently solve the problem.

1.2 Evolutionary Optimization

Evolutionary optimization is a class of black box optimization algorithms that mimics the biological process of optimization known as evolution. Evolutionary algorithms are based on evolutionary operators that model problem-specific processes in natural evolution, of which the most important are

1. crossover,
2. mutation, and
3. selection.

Basis of most evolutionary methods is a set of candidate solutions. Crossover combines the most promising characteristics of two or more solutions. Mutations adds random changes, while carefully balancing exploration and exploitation. Selection chooses the most promising candidate solutions in an iterative kind of way, alternately with recombination and mutation. Evolutionary algorithms have developed to strong optimization algorithms for difficult continuous optimization problems. An example for a hard optimization problem is Kursawe's function, which is defined as

$$f_{\mathrm{Kur}}(\mathbf{x}) = \sum_{i=1}^{N} \left(|x_i|^{0.8} + 5 \cdot \sin(x_i)^3 + 3.5828 \right). \tag{1.7}$$

The search for the minimum of Kursawe's function is difficult as it suffers from many local optima, see Fig. 1.2. The long line of research on evolutionary computation

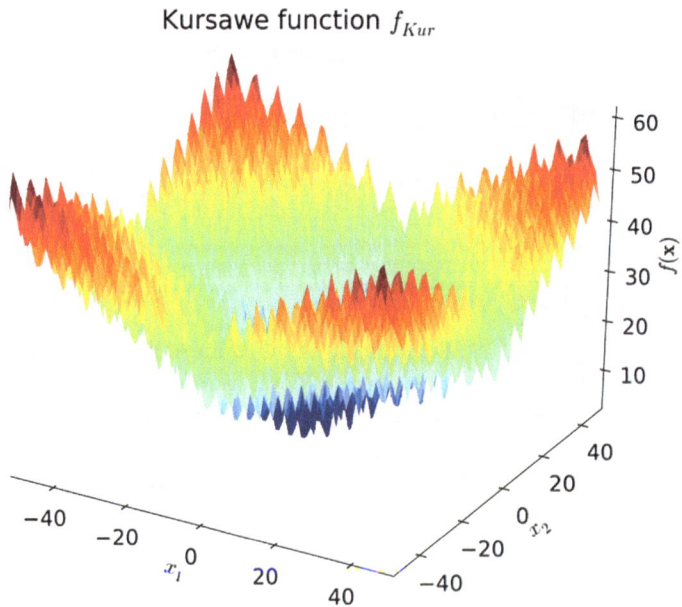

Fig. 1.2 Plot of the multimodal *Kursawe* function

was motivated by the goal of applying evolutionary algorithms to a wide range of problems. Applying evolutionary algorithms is comparatively easy, if the modeling of the problem is conducted carefully and appropriate representations and parameters are chosen. Chapter 2 will introduce evolutionary algorithms in more detail.

1.3 Machine Learning

The two problem classes supervised and unsupervised learning belong to the field of pattern recognition and machine learning. The idea of machine learning algorithms is to learn from observations. If one observes patterns $\mathbf{x}_i \in \mathbb{R}^q$ with $i = 1, \ldots, N$, and a label y_i (which may be a class) can be assigned to each pattern, the set of labeled patterns $(\mathbf{x}_1, y_1), \ldots, (\mathbf{x}_N, y_N)$ can be used to train a functional model f. The label can be a discrete class label or a continuous value $y_i \in \mathbb{R}$ (later also a vector). As the true distribution of patterns and labels is typically not known, the search for f can be performed by minimizing the empirical risk

$$E_{emp}(f) = \frac{1}{N} \sum_{i=1}^{N} L(f(\mathbf{x}_i), y_i) \tag{1.8}$$

based on the available observations with loss function $L(\cdot)$ measuring the deviations between predictions $f(\mathbf{x}_i)$ and labels y_i. This problem is also known as model selection. The optimal result is guaranteed, if the search takes place in the set of all functions $\mathcal{F}$

$$f^* = \arg \min_{f \in \mathcal{F}} E_{emp}(f). \tag{1.9}$$

In practice, it is not reasonable to search in the whole set $\mathcal{F}$. Instead, it is reasonable to choose a certain method corresponding to a function subset $F \subset \mathcal{F}$ and to optimize its free parameters w.r.t. the empirical risk resulting in model f. If the function space F is large, overfitting may occur, and it is a reasonable approach to restrict it by penalizing the complexity of model f with a regularizer, which is often a functional norm $\|f\|$. Then, the objective becomes to minimize the regularized risk

$$E_{reg}(f, \lambda) = E_{emp}(f) + \lambda \|f\|, \tag{1.10}$$

where $\lambda \in \mathbb{R}^+$ is a regularization parameter balancing between empirical risk minimization and smoothness of the function.

A comparatively simple classifier is K-nearest neighbor (KNN) classification [1]. For an unknown pattern $\mathbf{x}_j$, it assigns the class label of the K-closest patterns in data space. For this sake, a distance measure has to be defined in the space of patterns. In $\mathbb{R}^q$, it is reasonable to employ the Minkowski metric (p-norm)

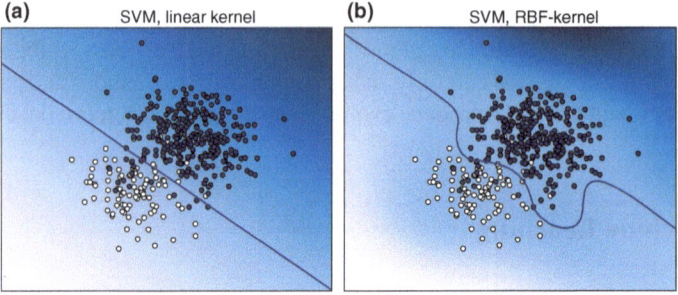

Fig. 1.3 Illustration of SVM classification on two overlapping Gaussian data clouds **a** with linear kernel and **b** with RBF-kernel

$$\delta(\mathbf{x}_j, \mathbf{x}_k) = \left(\sum_{i=1}^{q} |(x_i)_j - (x_i)_k|^p \right)^{1/p}, \tag{1.11}$$

which corresponds to the Euclidean distance for $p = 2$. In other solution spaces, adequate distance functions have to be chosen, e.g., the Hamming distance in $\mathbb{B}^N$. With the help of the distance function, it is now possible to define KNN

$$f(\mathbf{x}') = \begin{cases} 1 & \text{if } \sum_{i \in \mathfrak{N}_K(\mathbf{x}')} y_i > 0 \\ -1 & \text{if } \sum_{i \in \mathfrak{N}_K(\mathbf{x}')} y_i \leq 0, \end{cases} \tag{1.12}$$

with set $\mathfrak{N}_K(\mathbf{x}')$ containing the indices of the K-nearest neighbors of $\mathbf{x}'$ and neighborhood size K. The choice of K defines how *local* KNN is. For $K = 1$, little neighborhoods arise in regions, where patterns from different classes are scattered. For larger neighborhood sizes, patterns with labels in the minority are ignored. Further prominent classification methods are decision tress like ID3 [2], backpropagation networks [3, 4], and support vector machines (SVMs) [5, 6]. Figure 1.3 shows the SVM learning results for the classification of two overlapping Gaussian data clouds: (a) with a linear kernel and (b) with an RBF-kernel. The linear kernel SVM has chosen the separating decision boundary that maximizes the margin, i.e., the distance to the closest patterns of both classes. The RBF-kernel learns a non-linear decision boundary that is able to better separate both classes using a feature space. For a detailed introduction to SVMs, we refer the reader to Bishop [7] and Hastie et al. [8].

The question arises how to choose the parameters of supervised learning methods, e.g., kernel parameters of an SVM and neighborhood sizes of KNN. Various techniques like cross-validation can be used to choose the best model. Cross-validation divides the set of observed patterns into training and validation sets and successively computes the error w.r.t. different settings to avoid overfitting. Outlier or novelty detection is a special variant of supervised learning. The task is to learn an estimator of patterns with given labels and to let this classifier determine, if novel patterns belong to the same distribution, or if they can be classified as *outliers*.

Simulation models can be expensive. Optimization procedures in design processes may require a large number of function (response) evaluations. Often, due to complex relationships between models, analytic relationships cannot be determined leading to the black box optimization scenario. A significant reduction of computation effort (e.g., spent on complex simulation models) can be achieved employing meta-modeling techniques. Meta-models are machine learning models that are used as surrogates of the real simulation model. Based on past response evaluations, a statistical model is built that serves as basis for response evaluation estimates. This idea is related to the standard supervised learning scenario.

1.4 Hybrid Strategies

To overcome algorithmic shortcomings and achieve synergetic effects, hybridization of different methods can be an effective strategy. In recent years, a lot of research contributions concentrated on hybrid solution strategies. Many stem from the intersection of evolution strategies and machine learning, and most led to an improvement in comparison to the native strategies. But what are the common driving forces of the success of hybrid strategies?

The *no-free-lunch theorem* of Wolpert and Macready [9] states that there is no optimal algorithm for every problem. Algorithms must be tailored to special problem instances. A powerful strategy is to exploit the abilities of more than one algorithm. Two main principles hybrid strategies have in common are:

- Committee of experts: The combination of predictions in classification and the exchange of successful candidate solutions in optimization improves both classifiers and optimization techniques. The combination of results of more than one algorithm may lead to an improvement, instead of relying on a single expert.
- Informing search: Search can be improved by incorporating as much information about the solution space as possible. For example, in evolutionary computation candidate solutions should be initialized with a good guess of an optimal solution. Genetic operators should be employed that are tailored to the solution space. Machine learning and pattern recognition can deliver useful information that can hardly be modeled with genetic operators.

On the one hand machine learning methods turn out to be advantageous in informing and accelerating black box search. On the other hand, evolutionary and swarm-based search has been employed to optimize machine learning problems. Also in machine learning, hybridization has proven to be an effective way to improve the robustness of classifiers.

Algorithms in optimization can be divided into two categories: exact techniques and heuristics. Exact algorithms find local optimal solutions in guaranteed time, but the computational efficiency deteriorates significantly with the problem dimension. Heuristics and meta-heuristics that are design patterns for heuristics usually approximate the solution on the basis of stochastic components, but do not guarantee to

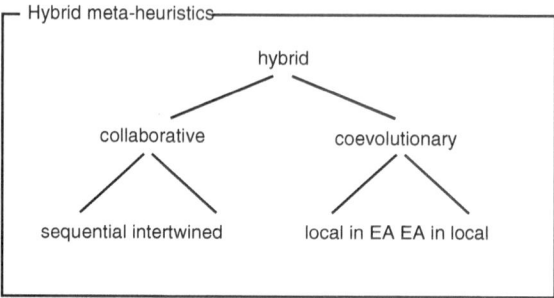

Fig. 1.4 Survey of hybridization strategies [10]. Hybrids can be divided into collaborative approaches that run successively (sequentially or intertwined) and coevolutionary approaches that employ further algorithms in each iteration (e.g., local search methods embedded in evolutionary optimizers or vice versa)

find the optimum in every case. However, their runtime on large problem instances is often more acceptable. The hybridization of meta-heuristics and local search methods is motivated by the combination of the advantages of the exact and the heuristic world.

An important design decision for hybrid techniques is the way of information exchange between their components. In which order shall the components work together, which information is shared and when? For a systematic overview, Talbi [11] and Raidl [12] introduced a taxonomy of hybrid meta-heuristics. Figure 1.4 shows a similar view on hybrid meta-heuristics based on their taxonomy. Hybrids can be classified into collaborative techniques that work successively or intertwined. A *sequential* hybrid employs a simple successive execution of two or more algorithmic components. The main idea is: A stochastic method *preoptimizes* coarsely, while the local search performs fine-tuning and approximates local optima. The intertwined collaborative hybrid is alternately running various optimizers. The coevolutionary hybrids represent the other branch of the taxonomy and are nested approaches. Typically, a local search method is embedded into an evolutionary optimizer. In each iteration, the local method optimizes the offspring solutions until a predefined termination condition is fulfilled. Information is passed alternately between the components in the concurrent approach. The local search method can use a separate termination condition that can be specified by the embedding optimizer. The alternative, i.e., to integrate evolutionary optimization into a local optimizer is rather unusual.

1.5 Overview of Chapters

The objective of this work is to offer extensions of evolutionary optimization and learning methods for special problem classes with an emphasis on continuous optimization. Most of the methodological advancements are based on hybridizations. This section gives an overview of the chapters.

Chapter 2

Chapter 2 gives an introduction to evolution strategies (ES). ES are nature-inspired meta-heuristics for optimization problems. The idea of translating evolutionary principles into an algorithmic framework for optimization has a long history. First, I introduce the classic evolution strategy, i.e., the $(\mu \overset{+}{,} \lambda)$-ES, which is the most famous evolution strategy variant employing intermediate or dominant recombination and Gaussian mutation. The self-adaptive control of step sizes allows to adapt to local solution space characteristics. The covariance matrix self-adaptation evolution strategy (CMSA-ES) by Beyer and Sendhoff [13] will also be introduced. An experimental evaluation of the CMSA-ES on a set of numerical black box functions completes Chap. 2 and will guide as reference for the later chapters.

Chapter 3

The success of many optimization algorithms depends on proper parameter choices. Although there are lot of studies on parameter control and specifically on mutation rates, there is no broad and comprehensive comparison of parameter control techniques that allows to give recommendations for certain scenarios. In Chap. 3, a comprehensive comparison of tuning and control techniques of continuous mutation rates employing the same algorithmic setting on a simple discrete unimodal problem is presented. After an analysis of various mutation rates for a $(1 + 1)$-EA on *OneMax*, I compare meta-evolution to Rechenberg's 1/5th rule and self-adaptation.

Chapter 4

In Chap. 4, I introduce an adaptive penalty function oriented to Rechenberg's 1/5th success rule to handle constraints. If less than 1/5th of the candidate population is feasible, the penalty is increased, otherwise, it is decreased. Experiments on the tangent problem demonstrate that this simple strategy leads to very successful results for the high-dimensional constrained sphere function. I accelerate the approach with two regression meta-models, one for the constraint and one for the fitness function.

Chapter 5

Direct search method like Powell's method are efficient black box optimization strategies for unimodal optimization problems. In case of multimodal optimization problems, the hybridization with ES turns out to be a fruitful strategy. In Chap. 5, we employ Powell's method as local search method in an ES-based global optimization process. A step size control strategy will be introduced that allows to search in multimodal solution spaces.

Chapter 6

In practical optimization, often two or more conflictive objectives have to be optimized at the same time. In Chap. 6, I present an approach that allows to select Pareto optimal solutions a posteriori based on reference lines in objective space. Reference lines define regions of interest in objective space that are basis of the selection of Pareto optimal solutions.

Chapter 7

The Nadaraya-Watson estimator, also known as kernel regression, is a successful density-based regression technique. Densities are measured with kernel functions that depend on bandwidth parameters. The choice of appropriate kernel parameters is an important problem in machine learning. In Chap. 7, an evolutionary kernel shape optimizer for kernel regression is presented. The approach is based on parameterized kernel density functions, leave-one-out cross-validation, and the CMSA-ES as optimization engine. A comparison to grid search shows that evolutionary search is an effective alternative for kernel parameter optimization.

Chapter 8

Learning mappings from high-dimensional data spaces to low-dimensional latent spaces is usually an computationally expensive optimization problem. Chapter 8 presents an iterative approach that constructs a solution using particle swarm optimization based steps in each iteration. Based on the framework of unsupervised regression, KNN regression is used for the dimensionality reduction mapping.

Appendix

The appendix gives an overview of optimization test problems used in this work that are typical test problems from evolutionary optimization literature. Furthermore, the appendix presents the machine learning data sets of the corresponding evolutionary learning chapters.

1.6 Preliminary Work

This work is a collection of heuristic extensions for ES to handle various classes of black box optimization and learning problems. Parts of this work are based on preliminary *peer-reviewed* publications of original research articles in international conferences and journals. The work provides a consistent view on the past research activities and highlights important preliminary work.

- Chapter 2 is based on a book chapter that gives an introduction to derivative-free optimization [14], complemented by an experimental analysis of the CMSA-ES.
- Chapter 3 is based on a publication on the *German Conference on Artificial Intelligence (KI) 2013* [15].
- The Rechenberg penalty function of Chap. 4 has been introduced on the *Congress on Evolutionary Computation (CEC) 2013* in Cancun, Mexico.
- An article on the Powell ES of Chap. 5 has been introduced in Springer's *Memetic Computing Journal* in 2009 [10, 16].
- The rake selection approach of Chap. 6 has been introduced on the *German Conference on Artificial Intelligence (KI) 2009* [17].
- The kernel regression approach of Chap. 7 is partly based on an article that has been published in the *Expert Systems and Applications Journal* in 2010 [18] with new experimental results.
- The particle swarm embedding algorithm of Chap. 8 has been introduced on the *Ant Colony Optimization and Swarm Intelligence (ANTS) 2012*.

This work is a cumulative and consistent depiction of the published research results, presenting various extended results and descriptions. The remainder of this work will be written in a scientific style with the use of "we" rather than "I".

References

1. N. Bhatia, Vandana, Survey of nearest neighbor techniques. CoRR, abs/1007.0085, (2010)
2. T. Mitchell, *Machine Learning* (McGraw Hill, Maidenhead, 1997)
3. R. Rojas, *Neural Networks - A Systematic Introduction* (Springer, Berlin, 1996)
4. D. Rumelhart, G. Hintont, R. Williams, Learning representations by backpropagating errors. Nature **323**(6088), 533–536 (1986)
5. B. Schölkopf, A.J. Smola, *Learning with Kernels: Support Vector Machines, Regularization, Optimization, and Beyond* (MIT Press, Cambridge, 2001)
6. J.A.K. Suykens, J. Vandewalle, Least squares support vector machine classifiers. Neural Process. Lett. **9**(3), 293–300 (1999)
7. C.M. Bishop, *Pattern Recognition and Machine Learning (Information Science and Statistics)* (Springer, New York, 2007)
8. T. Hastie, R. Tibshirani, J. Friedman, *The Elements of Statistical Learning* (Springer, Berlin, 2009)
9. D.H. Wolpert, W.G. Macready, No Free Lunch Theorems for Optimization. IEEE Trans. Evol. Comput. **1**(1), 67–82 (1997)
10. O. Kramer, Iterated local search with Powell's method: a memetic algorithm for continuous global optimization. Memetic Comput. **2**(1), 69–83 (2010)
11. E.-G. Talbi, A taxonomy of hybrid metaheuristics. J. Heuristics **8**(5), 541–564 (2002)
12. G.R. Raidl, *A unified view on hybrid metaheuristics. in Hybrid Metaheuristics (HM)* (Springer, Gran Canaria, 2006), pp. 1–12
13. H.-G. Beyer, B. Sendhoff, Covariance matrix adaptation revisited - the CMSA evolution strategy, in Proceedings of the 10th Conference on Parallel Problem Solving from Nature (PPSN), 2008, pp. 123–132

14. O. Kramer, D. E. Ciaurri, S. Koziel, Derivative-Free Optimization, in Computational Optimization and Applications in Engineering and Industry, Studies in Computational Intelligence, Springer, 2011, pp. 61–83
15. O. Kramer, On Mutation Rate Tuning and Control for the (1+1)-EA, in International Conference on Artificial, Artificial Intelligence, 2013, pp. 98–105
16. O. Kramer. Fast black box optimization: iterated local search and the strategy of powell. in International Conference on Genetic and Evolutionary Methods (GEM), CSREA Press, 2009
17. O. Kramer, P. Koch, Rake selection: A novel evolutionary multi-objective optimization algorithm. in Proceedings of the German Annual Conference on Artificial Intelligence (KI), Springer, Berlin, pp. 177–184
18. O. Kramer, F. Gieseke, Evolutionary kernel density regression. Expert Syst. Appl. **39**(10), 9246–9254 (2012)

Chapter 2
Evolution Strategies

2.1 Introduction

Many real-world problems are multimodal, which renders an optimization problem difficult to solve. Local search methods, i.e., methods that greedily improve solutions based on search in the neighborhood of a solution, often only find an arbitrary local optimum that is not guaranteed to be the global one. The most successful methods in global optimization are based on stochastic components, as they allow to escape from local optima and overcome premature stagnation. A famous class of global optimization methods are evolution strategies that are successful in real-valued solution spaces. Evolution strategies belong to the most famous evolutionary methods for black box optimization, i.e., for optimization scenarios, where no functional expressions are explicitly given and no derivatives can be computed. In the course of this work, evolution strategies will play an important role. They are oriented to the biological principle of evolution [1] and can serve as an excellent starting point to methods in learning and optimization. They are based on three main mechanisms that are translated into evolutionary operators:

1. recombination,
2. mutation, and
3. selection.

First, we define an optimization problem formally. Let $f : S \rightarrow \mathbb{R}$ be the fitness function to be minimized in the space of solutions S. The problems we consider in this work are minimization problems unless explicitly stated. *High* fitness means *low* fitness values. The task is to find an element $\mathbf{x}^* \in S$ such that $f(\mathbf{x}^*) \leq f(\mathbf{x})$ for all $\mathbf{x} \in S$. A desirable property of an optimization method is to find the optimum $\mathbf{x}^*$ with fitness f^* within a finite and preferably low number of function evaluations. In most parts of this work, we consider continuous optimization problems, i.e., the solution space $S = \mathbb{R}^N$. Problem f can be an arbitrary optimization problem, e.g., a civil engineering system like a simulation or a mathematical model.

O. Kramer, *A Brief Introduction to Continuous Evolutionary Optimization*,
SpringerBriefs in Computational Intelligence,
DOI: 10.1007/978-3-319-03422-5_2, © The Author(s) 2014

2.2 Evolutionary Algorithms

If derivatives are available, Newton methods and variants are recommendable algorithmic choices. From this class of methods, the Broyden-Fletcher-Goldfarb-Shanno (BFGS) algorithm belongs to the state-of-the-art techniques [2]. This work concentrates on black box optimization problems. In black box optimization, the problem does not have to fulfill any assumptions or limiting properties. For such general optimization scenarios, evolutionary methods are a good choice. Evolutionary algorithms (EAs) belong to the class of stochastic derivative-free optimization methods. Their biological motivation has made them very popular. After decades of research, a long history of applications and theoretical investigations have proven their success.

In Germany, the history of evolutionary computation began with evolution strategies, which were developed by Rechenberg and Schwefel in the sixties and seventies of the last century in Berlin [3–5]. At the same time, John Holland introduced the evolutionary computation concept in the United States known as *genetic algorithms* [6]. Today, advanced mutation operators, step size mechanisms, and methods to adapt the covariance matrix like the CMA-ES [7] have made them one of the most successful optimizers in derivative-free global optimization.

Many methods have been presented in evolutionary continuous optimization like the work by Deb et al. [8], who developed a generic parent-centric crossover operator, and a steady-state, elite-preserving population-alteration model. Herrera et al. [9, 10] proposed to apply a two-loop EA with adaptive control of mutation sizes. The algorithm adjusts the step size of an inner EA and a restart control of a mutation operator in the outer loop. Differential evolution (DE) is another branch of evolutionary methods for continuous optimization. Price et al. [11] give an introductory survey to DE. Qin et al. [12] proposed an adaptive DE that learns operator selection and associated control parameter values. The learning process is based on previously generated successful solutions. Particle swarm optimization (PSO) is a famous methodology that concentrates on continuous global optimization [13, 14]. PSO is inspired by the movement of swarms in nature, e.g., fish schools or flocks of birds. It simulates the movement of candidate solutions using flocking-like equations with locations and velocities. A learning strategy variant has been proposed by Liang et al. [15], who uses all particles' past best information to update the particle history. A PSO-like algorithm will be employed in Chap. 8.

Evolutionary search is based on a set $\mathcal{P} = \{\mathbf{x}_1, \ldots, \mathbf{x}_\mu\}$ of parental and a set $\mathcal{P}' = \{\mathbf{x}_1, \ldots, \mathbf{x}_\lambda\}$ of offspring candidate solutions, also called individuals. The individuals are iteratively subject to random changes and selection of the best solutions. Algorithm 1 shows the pseudocode of a general evolutionary algorithm. The optimization process consists of three main steps:

1. The recombination operator selects ρ parents and combines their parts to λ new solutions.
2. The mutation operator adds random changes (e.g. noise) to the preliminary candidate solution. The quality of the individuals in solving the optimization problem

---

Algorithm 1 Evolutionary Algorithm

1: initialize solutions $\mathbf{x}_1, \ldots, \mathbf{x}_\mu \in \mathcal{P}$
2: **repeat**
3:    **for** $i = 1$ **to** λ **do**
4:       select ρ parents from $\mathcal{P}$
5:       create $\mathbf{x}_i$ by recombination
6:       mutate $\mathbf{x}_i$
7:       evaluate $\mathbf{x}_i \rightarrow f(\mathbf{x}_i)$
8:       add $\mathbf{x}_i$ to $\mathcal{P}'$
9:    **end for**
10:   select μ parents from $\mathcal{P}' \rightarrow \mathcal{P}$
11: **until** termination condition

---

is called fitness. The fitness of the new offspring solution is evaluated on fitness function f. All individuals of a generation are put into offspring population $\mathcal{P}'$.

3. Then, μ individuals are selected and constitute the novel parental population $\mathcal{P}$ of the following generation.

The process is repeated until a termination condition is reached. Typical termination conditions are defined via fitness values or via an upper bound on the number of generations.

In the following, we will give a short survey of evolutionary operators and go deeper into evolution strategies that have proven well in practical optimization scenarios. The evolution strategy operators intermediate and dominant recombination as well as Gaussian mutation are introduced.

2.3 Recombination

In biological systems, recombination, also known as crossover, mixes the genetic material of two parents. Most evolutionary algorithms also make use of a recombination operator and combine the information of two or more individuals $\mathbf{x}_1, \ldots, \mathbf{x}_\rho$ to a new offspring solution. Hence, the offspring carries parts of the genetic material of its parents. Many recombination operators are restricted to two parents, but also multi-parent recombination variants have been proposed in the past that combine information of ρ parents. The use of recombination is discussed controversially within the building block hypothesis by Goldberg [16, 17]. The building block hypothesis assumes that good substrings of the solutions called building blocks of different parents are combined, and their number increases. The *good genes* are spread over the population in the course of the evolutionary process.

Typical recombination operators for continuous representations are dominant and intermediate recombination. Dominant recombination randomly combines the genes of all parents. Dominant recombination with ρ parents $(\mathbf{x})_1, \ldots, (\mathbf{x})_\rho \in \mathbb{R}^N$ creates the offspring vector $\mathbf{x}' = (x'_1, \ldots, x'_N)^T$ by randomly choosing the i-th component

$$x'_i = (x_i)_k, \quad k \in \text{random}\{1, \ldots, \rho\}. \tag{2.1}$$

Intermediate recombination is appropriate for numerical solution spaces. Given ρ parents $\mathbf{x}_1, \ldots, \mathbf{x}_\rho$ each component of the offspring vector $\mathbf{x}'$ is the arithmetic mean of the components of all ρ parents

$$x_i' = \frac{1}{\rho} \sum_{k=1}^{\rho} (x_i)_k. \tag{2.2}$$

The characteristics of offspring solutions lie between their parents. Integer representations may require rounding procedures for generating valid solutions.

2.4 Mutation

Mutation is the second main source of evolutionary changes. According to Beyer and Schwefel [3], a mutation operator is supposed to fulfill three conditions. First, from each point in the solution space each other point must be reachable. Second, in unconstrained solution spaces a bias is disadvantageous, because the direction to the optimum is unknown, and third, the mutation strength should be adjustable, in order to adapt exploration and exploitation to local solution space conditions.

In the following, we concentrate on the famous Gaussian mutation operator for optimization in $\mathbb{R}^N$. Solutions are vectors of real values $\mathbf{x} = (x_1, \ldots, x_N)^T \in \mathbb{R}^N$. Random numbers based on the Gaussian distribution $\mathcal{N}(0, 1)$ fulfill these conditions in continuous domains.[1] With the Gaussian distribution, many natural and artificial processes can be described. The idea is to mutate each individual applying the mutation operator

$$\mathbf{x}' = \mathbf{x} + \mathbf{z}, \tag{2.3}$$

with a mutation vector $\mathbf{z} \in \mathbb{R}^N$ based on sampling from the Gaussian distribution

$$\mathbf{z} \sim \mathcal{N}(\mathbf{0}, \sigma^2 \mathbf{I}) = (\mathcal{N}(0, \sigma^2), \ldots, \mathcal{N}(0, \sigma^2))^T \sim \sigma \mathcal{N}(\mathbf{0}, \mathbf{I}) \tag{2.4}$$

with identity matrix $\mathbf{I}$. The standard deviation σ plays the role of the mutation strength and is also known as step size. The isotropic Gaussian mutation with only one step size uses the same standard deviation for each component x_i. Of course, the step size σ can be kept constant, but the convergence to the optimum can be improved by adapting σ according to local solution space characteristics. In case of high success rates, i.e., a large number of offspring solutions being better than their parents, big step sizes are advantageous, in order to explore the solution space as fast as possible. This is often reasonable at the beginning of the search. In case of low success rates, smaller step sizes are appropriate. This is often adequate in later phases of the search during convergence to the optimum, i.e., when approximating solutions should not

---

[1] $\mathcal{N}(m, \sigma^2)$ represents a randomly drawn Gaussian distributed number with expectation value m and standard deviation σ.

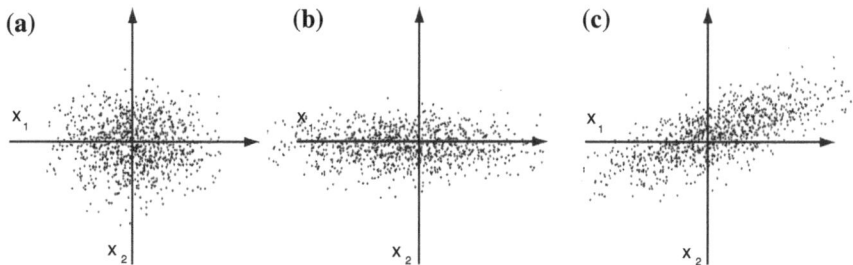

Fig. 2.1 Gaussian mutation: **a** isotropic Gaussian mutation employs one step size σ for each dimension, **b** multivariate Gaussian mutation allows independent step sizes in each dimension, and **c** correlated mutation allows a rotation of the mutation ellipsoid, (**a**) isotropic, (**b**) multivariate, (**c**) correlated

be destroyed. An example for an adaptive control of step sizes is the 1/5-th success rule by Rechenberg [4] that increases the step sizes, if the success rate is over 1/5-th, and decreases it, if the success rate is lower.

Isotropic Gaussian mutation can be extended to multivariate Gaussian mutation by allowing independent scalings of the components

$$\mathcal{N}(\mathbf{0}, \mathbf{D}^2) = (\mathcal{N}(0, \sigma_1^2), \ldots, \mathcal{N}(0, \sigma_N^2))^T \sim \mathbf{D}\mathcal{N}(\mathbf{0}, \mathbf{I}). \tag{2.5}$$

This multivariate mutation employs N degrees of freedom that are saved in a diagonal matrix $\mathbf{D} = \text{diag}(\sigma_1, \ldots, \sigma_N)$ corresponding to a step size vector for the independent scalings.

Figure 2.1 illustrates the differences between (a) isotropic Gaussian mutation and (b) multivariate Gaussian mutation. The multivariate variant allows the development of a Gaussian ellipsoid that flexibly adapts to local solution space characteristics. Even more flexibility, i.e., $N(N-1)/2$ degrees of freedom, allows the correlated mutation presented by Schwefel [18]

$$\mathcal{N}(\mathbf{0}, \mathbf{C}) = \sqrt{\mathbf{C}}\mathcal{N}(\mathbf{0}, \mathbf{I}) \tag{2.6}$$

with covariance matrix $\mathbf{C}$, which contains the covariances describing the multivariate normal distribution. The components are correlated, see Fig. 2.1c. The *square root*, or Cholesky decomposition, $\sqrt{\mathbf{C}}$ of the covariance matrix $\mathbf{C}$ corresponds to a rotation matrix for the mutation ellipsoid axes. The question arises, how to control the mutation ellipsoid rotation. Instead of a rotation matrix $N(N-1)/2$ angles can be used. In practical optimization, these angles are often controlled self-adaptively [19]. Also the CMA-ES and variants [7, 20] are based on an automatic alignment of the coordinate system (cf. Sect. 2.7).

A step towards the acceleration of the step size control is σ-self-adaptation. Before the application of the mutation operator (cf. Eqs. 2.3 and 2.4), the log-normal mutation operator for step sizes σ and step size vectors $(\sigma_1, \ldots, \sigma_N)^T$ is applied. The log-normal mutation operator has been proposed by Schwefel [18] and has become famous for self-adaptation in continuous solution spaces. It is defined as

$$\sigma' = \sigma \cdot e^{\tau \cdot \mathcal{N}(0,1)}. \tag{2.7}$$

The problem-dependent learning rate τ has to be chosen adequately. For the mutation strengths of evolution strategies on the continuous *Sphere* model, theoretical investigations [3] lead to the optimal setting

$$\tau = \frac{1}{\sqrt{N}}, \tag{2.8}$$

which may not be optimal for other problems, and further parameter tuning is recommended. Strategy parameter σ cannot become negative and scales logarithmically between values close to 0 and infinity.[2] A more flexible approach is to mutate each of the N dimensions independently

$$\sigma' = e^{(\tau_0 \mathcal{N}(0,1))} \cdot \left(\sigma_1 e^{(\tau_1 \mathcal{N}(0,1))}, \ldots, \sigma_N e^{(\tau_1 \mathcal{N}(0,1))} \right), \tag{2.9}$$

with

$$\tau_0 = \frac{c}{\sqrt{2 \cdot N}}, \tag{2.10}$$

and

$$\tau_1 = \frac{c}{\sqrt{2\sqrt{N}}}. \tag{2.11}$$

Setting parameter $c = 1$ is a recommendable choice. *Kursawe* [21] analyzed parameters τ_0 and τ_1 using a nested evolution strategy on various test problems. His analysis shows that the choice of mutation parameters is problem-depended, and general recommendations are difficult to give.

The EA performs the search in two spaces: the objective and the strategy parameter space. Strategy parameters influence the genetic operators of the objective variable space, in this case the step sizes of the Gaussian mutation. The optimal settings may vary depending on the location of the solution in the fitness landscape. Only the objective variables define the solution and have an impact on the fitness. Strategy parameters have to take part in the evolutionary process to evolve them dynamically during the optimization process.

2.5 Selection

The counterpart of the variation operators mutation and recombination is selection. Selection gives the evolutionary search a direction. Based on their fitness, a subset of the population is selected, while the worst individuals are rejected. In the evolutionary framework, the selection operator can be employed at two steps. Mating selection

---

[2] i.e., high values w.r.t. the data structure.

selects individuals for the recombination operator. In nature, the attraction of sexual partners as well as cultural aspects influence the mating selection process. The second famous selection operator is survivor selection corresponding to the Darwinian principle of *survival of the fittest*. Only the individuals selected by survivor selection are allowed to inherit their genetic material to the following generation. The probability of a solution to be selected is also known as selection pressure.

Evolution strategies usually do not employ a competitive selection operator for mating selection. Instead, parental solutions are randomly drawn from the set of candidate solutions. But for survivor selection, the elitist selection strategies comma and plus selection are used. They choose the μ-best solutions as basis for the parental population of the following generation. Both operators, plus and comma selection, can easily be implemented by sorting the population w.r.t. the individuals' fitness. Plus selection selects the μ-best solutions from the union $\mathcal{P} \cup \mathcal{P}'$ of the last parental population $\mathcal{P}$ and the current offspring population $\mathcal{P}'$, and is denoted by $(\mu + \lambda)$-ES. In contrast, comma selection in a (μ, λ)-ES selects exclusively from the offspring population, neglecting the parental population, even if the parents have a superior fitness. Forgetting superior solutions may seem to be disadvantageous. But potentially good solutions may turn out to be local optima, and the evolutionary process may fail to leave them without the ability to forget.

2.6 Particle Swarm Optimization

Similar to evolutionary algorithms, PSO is a population approach with stochastic components. Introduced by Kennedy and Eberhart [13], it is inspired by the movement of natural swarms and flocks. The algorithm utilizes particles with a position $\mathbf{x}$ that corresponds to the optimization variables and a velocity $\mathbf{v}$, which is similar to the mutation strengths in evolutionary computation. The principle of PSO is based on the idea that the particles move in solution space influencing each other with stochastic changes, while previous successful solutions act as attractors. Figure 2.2 illustrates the PSO concept in $N = 2$ dimensions, while Algorithm 2 shows the PSO algorithm in pseudocode.

In each iteration the position of particle $\mathbf{x}$ is updated by adding a velocity $\hat{\mathbf{v}}$

Fig. 2.2 Illustration of PSO concept

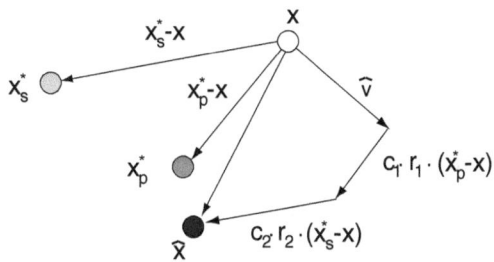

Algorithm 2 Particle Swarm Optimization Algorithm

1: initialize parameters, and particles
2: **repeat**
3:    **for** $i = 1$ **to** κ **do**
4:        compute $\mathbf{x}_p^*$, and $\mathbf{x}_s^*$
5:        update velocity $\hat{\mathbf{v}}$
6:        update position $\hat{\mathbf{x}}$
7:        compute fitness $f(\hat{\mathbf{x}})$
8:    **end for**
9: **until** termination condition

$$\hat{\mathbf{x}} = \mathbf{x} + \hat{\mathbf{v}}, \tag{2.12}$$

which is updated as follows

$$\hat{\mathbf{v}} = \mathbf{v} + c_1 r_1 (\mathbf{x}_p^* - \mathbf{x}) + c_2 r_2 (\mathbf{x}_s^* - \mathbf{x}), \tag{2.13}$$

where $\mathbf{x}_p^*$, and $\mathbf{x}_s^*$ denote the best previous positions of the particle, and of the swarm, respectively. The weights $c_1, c_2 \in [0, 1]$ are acceleration coefficients that determine the bias of the particle towards its own, and the swarm history. The recommendation given by Kennedy and Eberhart is to set both parameters to $c_1 = c_2 = 0.5$ [13]. The random components r_1, and r_2 are uniformly drawn from the interval $[0, 1]$, and can be used to control exploitation and exploration of the solution space.

2.7 Covariance Matrix Adaptation Evolution Strategies

In the following, we introduce an algorithm from the family of covariance matrix adaptation evolution strategies (CMA-ES). The covariance matrix self-adaptation evolution strategy (CMSA-ES) by Beyer and Sendhoff [20] is the historically latest covariance matrix adaptation-based strategy, but is a variant that reflects well the main idea of the family of CMA-ES. The basic idea is to align the coordinate system for the mutation operator to the distribution of the selected solutions in each generation. The aligned coordinate system guarantees that mutations in the following generation are similar to the best of the previous generation. The CMSA-ES is a variant of the famous CMA-ES by Hansen and Ostermeier [7] with an emphasis on self-adaptation.

The CMSA-ES is based on a self-adaptive step control of step sizes, similar to the $(\mu \overset{+}{,} \lambda)$-ES introduced in the previous section. After initialization, λ candidate solutions $\mathbf{x}_1, \ldots, \mathbf{x}_\lambda$ are generated. With the help of the global self-adaptive, N-dimensional step size $\hat{\sigma} = \frac{1}{\mu} \sum_{j=1}^{\mu} \sigma_{j:\lambda}$, which is the arithmetic mean of the step sizes from the μ-best solutions of λ offspring solutions[3] of the previous generation, each individual gets a log-normally mutated step size

---

[3] The index j denotes the index of the j-th ranked individual of the λ offspring individuals w.r.t. an increasing sorting based on fitness $f(\mathbf{x}_j)$.

$$\sigma_j = \hat{\sigma} \cdot e^{\tau_\sigma \mathcal{N}(0,1)}. \tag{2.14}$$

The main idea of the approach is to align the coordinate system by changing the coordinates $\mathbf{x}_j$ with the help of the current mean $\hat{\mathbf{x}}$ of the population and a covariance matrix $\mathbf{C}$ based on the best solutions and the past optimization process. From $\mathbf{C}$ the correlated random directions $\mathbf{s}_j$ are generated by multiplication of the Cholesky decomposition $\sqrt{\mathbf{C}}$ with the standard normal vector $\mathcal{N}(\mathbf{0}, \mathbf{I})$

$$\mathbf{s}_j \sim \sqrt{\mathbf{C}} \mathcal{N}(\mathbf{0}, \mathbf{I}). \tag{2.15}$$

This random direction is scaled in length w.r.t. the self-adaptive step size σ_j

$$\mathbf{z}_j = \sigma_j \mathbf{s}_j. \tag{2.16}$$

The resulting vector $\mathbf{z}_j$ is added to the global parent $\hat{\mathbf{x}}$, i.e.

$$\mathbf{x}_j = \hat{\mathbf{x}} + \mathbf{z}_j. \tag{2.17}$$

Finally, fitness $f_j = f(\mathbf{x}_j)$ of solution $\mathbf{x}_j$ is evaluated. When λ offspring solutions have been generated, the μ-best solutions are selected and their components $\mathbf{z}_j$ and σ_j are recombined. Beyer and Sendhoff [20] apply global recombination, i.e., the arithmetic mean of each parameter is computed. The outer product $\mathbf{ss}^T$ of the search directions is an estimation of the covariance matrix of the best mutations and is computed for each of the μ-best solutions and averaged afterwards

$$\mathbf{S} = \frac{1}{\mu} \sum_{j=1}^{\mu} \mathbf{s}_{j:\lambda} \mathbf{s}_{j:\lambda}^T. \tag{2.18}$$

Last, the covariance matrix $\mathbf{C}$ is updated based on the current covariance matrix and the new estimation $\mathbf{S}$. The covariance matrix update

$$\mathbf{C} = \left(1 - \frac{1}{\tau_c}\right) \mathbf{C} + \frac{1}{\tau_c} \mathbf{S} \tag{2.19}$$

is a composition of the last covariance matrix $\mathbf{C}$ and the outer product of the search direction of the μ-best solutions balanced by learning parameter τ_c. Adapted in such a kind of way, sampling from a Gaussian distribution based on $\mathbf{C}$ increases the likelihood of successful steps. Beyer and Sendhoff recommend to set

$$\tau_c = \frac{N(N+1)}{2\mu} \tag{2.20}$$

for the learning parameter. All steps are repeated until a termination condition is satisfied. The CMSA-ES combines the self-adaptive step size control with a simul-

---

Algorithm 3 CMSA-ES

1: initialize solution $\hat{\mathbf{x}}$
2: **repeat**
3:   **for** $i = 1$ **to** λ **do**
4:     $\sigma_j \sim \hat{\sigma} \cdot e^{\tau_\sigma \mathcal{N}(0,1)}$
5:     $\mathbf{s}_j \sim \sqrt{\mathbf{C}} \mathcal{N}(\mathbf{0}, \mathbf{I})$
6:     $\mathbf{z}_j = \sigma_j \mathbf{s}_j$
7:     $\mathbf{x}_j = \hat{\mathbf{x}} + \mathbf{z}_j$
8:     $f_j = f(\mathbf{x}_j)$
9:   **end for**
10: sort population w.r.t. fitness $f_{j:\lambda}$
11: $\hat{\mathbf{z}} = \frac{1}{\mu} \sum_{j=1}^{\mu} \mathbf{z}_{j:\lambda}$
12: $\hat{\sigma} = \frac{1}{\mu} \sum_{j=1}^{\mu} \sigma_{j:\lambda}$
13: $\mathbf{S} = \frac{1}{\mu} \sum_{j=1}^{\mu} \mathbf{s}_{j:\lambda} \mathbf{s}_{j:\lambda}^T$
14: $\hat{\mathbf{x}} = \hat{\mathbf{x}} + \hat{\mathbf{z}}$
15: $\mathbf{C} = (1 - \frac{1}{\tau_c})\mathbf{C} + \frac{1}{\tau_c}\mathbf{S}$
16: **until** termination condition

---

taneous update of the covariance matrix. Algorithm 3 shows the pseudocode of the CMSA-ES. Initially, the covariance matrix $\mathbf{C}$ is chosen as the identity matrix $\mathbf{C} = \mathbf{I}$. The learning parameter τ_σ defines the mutation strength of the step sizes σ_j. For the *Sphere* problem, the optimal learning parameter is $\tau_\sigma = \frac{1}{\sqrt{2 \cdot N}}$ [3].

In the following, we present an experimental analysis of the CMSA-ES concentrating on typical test problems known in literature [22] (cf. Appendix A). We use the following performance measure. The experimental results show the number of fitness function evaluations until the optimum is reached with accuracy f_{stop}, i.e., if the difference between the best achieved fitness $f(\mathbf{x}')$ of the algorithm and fitness $f(\mathbf{x}^*)$ of the known optimum $\mathbf{x}^*$ is smaller than f_{stop}, i.e.,

$$|f(\mathbf{x}') - f(\mathbf{x}^*)| \leq f_{\text{stop}}. \tag{2.21}$$

This performance measure is focused on the convergence abilities of the approach. The figures of Table 2.1 show the best, median, worst, and mean (with standard deviation) number of generations until the termination condition has been reached. The termination condition is set to $f_{\text{stop}} = 10^{-10}$.

The results confirm that the CMSA-ES is a strong method for derivative-free multimodal optimization. It is able to find the optima of all test problems. In case of the unimodal problems *Sphere* and *Doublesum*, no restarts have been necessary. The performance comparison in the later chapters will allow a detailed interpretation.

Table 2.1 Experimental analysis of CMSA-ES with restarts

	Best	Median	Worst	Mean	Dev
$N = 10$					
f_{Sp}	2,120	2,195	2,350	2,204.5	7.0e1
f_{Dou}	2,280	2,355	2,490	2,358.3	6.2e1
f_{Ros}	7,060	10,550	18,080	11,292.0	4.1e3
f_{Ras}	36,360	90,540	203,120	103,456.0	5.7e4
f_{Gri}	2,150	4,375	13,090	5,579.7	4.1e3
f_{Kur}	10,780	21,960	81,370	29,670.9	22.0e3
$N = 30$					
f_{Sp}	5,684	5,880	6,118	5,896.8	1.4c2
f_{Dou}	7,770	8,092	8,302	8,075.2	1.6e2
f_{Ros}	45,976	51,681	109,984	58,595.6	1.9e4
f_{Ras}	360,990	699,846	721,224	576,511.6	1.7e5
f_{Gri}	6,370	6,755	17,374	8,764	4.4e3
f_{Kur}	55,244	89,138	138,670	93,518.6	37.4e3

2.8 Conclusions

Evolution strategies, in particular the covariance matrix adaptation variants, belong to the most successful evolutionary optimization algorithms for solving black box optimization problems. If no derivatives are given and no assumptions about the fitness function are available, the application of evolutionary algorithms is a recommendable undertaking. Theoretical results and a huge variety of applications have proven their success in the past. But the success of evolutionary search also depends on proper parameter settings before and during the search. We will concentrate on the parameter control problem in the next chapter. We have already introduced σ-self-adaptation as parameter control techniques for steps sizes in evolution strategies, which is based on evolutionary search in the space of step sizes. This mechanism has significantly contributed to the success of evolutionary optimization methods.

References

1. O. Kramer, D. E. Ciaurri, S. Koziel, Derivative-free optimization, in *Computational Optimization and Applications in Engineering and Industry, Studies in Computational Intelligence* (Springer, New York, 2011). pp. 61–83
2. J. Nocedal, S. J. Wright, *Numerical Optimization* (Springer, New York, 2000)
3. H.-G. Beyer, H.-P. Schwefel, Evolution strategies—a comprehensive introduction. Nat. Comput. **1**, 3–52 (2002)
4. I. Rechenberg, *Evolutionsstrategie: Optimierung technischer Systeme nach Prinzipien der biologischen Evolution* (Frommann-Holzboog, Stuttgart, 1973)
5. H.-P. Schwefel, *Numerische Optimierung von Computer-Modellen mittels der Evolutionsstrategie* (Birkhäuser, Basel, 1977)

6. J.H. Holland, *Adaptation in Natural and Artificial Systems*, 1st edn (MIT Press, Cambridge, 1992)
7. A. Ostermeier, A. Gawelczyk, N. Hansen, A derandomized approach to self adaptation of evolution strategies. Evol. Comput. **2**(4), 369–380 (1994)
8. K. Deb, A. Anand, D. Joshi, A computationally efficient evolutionary algorithm for real-parameter optimization. Evol. Comput. **10**(4), 371–395 (2002)
9. F. Herrera, M. Lozan, J.L. Verdegay, Tackling real-coded genetic algorithms: operators and tools for behavioural analysis. Artif. Intell. Rev. **12**, 265–319 (1998)
10. F. Herrera, M. Lozano, Two-loop real-coded genetic algorithms with adaptive control of mutation step sizes. Appl. Intell. **13**(3), 187–204 (2000)
11. K.V. Price, R.M. Storn, J.A. Lampinen, *Differential Evolution A Practical Approach to Global Optimization* (Springer, Natural Computing Series, New York, 2005)
12. A.K. Qin, V.L. Huang, P.N. Suganthan, Differential evolution algorithm with strategy adaptation for global numerical optimization. IEEE Trans. Evol. Comput. **13**(2), 398–417 (2009)
13. J. Kennedy, R. Eberhart, Particle swarm optimization, in *Proceedings of IEEE International Conference on Neural Networks* (1995). pp. 1942–1948
14. Y. Shi, R. Eberhart, A modified particle swarm optimizer, in *Proceedings of the International Conference on Evolutionary Computation* (1998). pp. 69–73
15. J.J. Liang, A.K. Qin, P.N. Suganthan, S. Baskar, Comprehensive learning particle swarm optimizer for global optimization of multimodal functions. IEEE Trans. Evol. Comput. **10**(3), 281–295 (2006)
16. D. Goldberg, *Genetic Algorithms in Search* (Optimization and Machine Learning. Addison-Wesley, Reading, 1989)
17. J.H. Holland, *Hidden Order: How Adaptation Builds Complexity* (Addison-Wesley, Reading, 1995)
18. H.-P. Schwefel, *Adaptive Mechanismen in der biologischen Evolution und ihr Einfluss auf die Evolutionsgeschwindigkeit* (Interner Bericht der Arbeitsgruppe Bionik und Evolutionstechnik am Institut für Mess- und Regelungstechnik, TU Berlin, July 1974)
19. O. Kramer, *Self-Adaptive Heuristics for Evolutionary Computation, Studies in Computational Intelligence* (Springer, Heidelberg, 2008)
20. H.-G. Beyer, B. Sendhoff, Covariance matrix adaptation revisited—the CMSA evolution strategy, in *Proceedings of the 10th Conference on Parallel Problem Solving from Nature* (*PPSN*, 2008) pp. 123–132
21. F. Kursawe, Grundlegende empirische Untersuchungen der Parameter von Evolutionsstrategien—Metastrategien. PhD thesis, University of Dortmund, 1999
22. P.N. Suganthan, N. Hansen, J.J. Liang, K. Deb, Y.p. Chen, A. Auger, S. Tiwari, Problem definitions and evaluation criteria for the CEC special Session on real-parameter optimization, Technical report, Nanyang Technological University, 2005

Chapter 3
Parameter Control

3.1 Introduction

Parameter control is an essential aspect of successful evolutionary search. Various parameter control and tuning methods have been proposed in the history of evolutionary computation, cf. Fig. 3.1 for a short taxonomy. The importance of parameter control has become famous for mutation rates. Mutation is a main source of evolutionary changes. Mutation rates control the magnitude of random changes of solutions. At the beginning of the history of evolutionary computation, researchers argued about proper settings. De Jong's [1] recommendation was the mutation strength $\sigma = 0.001$, Schaffer et al. [2] recommended the setting $0.005 \leq \sigma \leq 0.01$, and Grefenstette [3] $\sigma = 0.01$. Mühlenbein [4] suggested to set the mutation probability to $\sigma = 1/N$ depending on the length N of the representation. But early, the idea appeared to control the mutation rate during the optimization run, as the optimal rate might change during the optimization process, and different rates are reasonable for different problems. Objective of this chapter is to compare the parameter tuning and control techniques of a simple evolutionary algorithm (EA) on a simple function, i.e., *OneMax*, to allow insights into the interplay of mutation rates and parameter control mechanisms. *OneMax* is a maximization problem defined on $\{0, 1\}^N \rightarrow \mathbb{N}$ that counts the number of ones in bit string $\mathbf{x}$

$$OneMax(\mathbf{x}) = \sum_{i=1}^{N} x_i. \tag{3.1}$$

The optimal solution is $\mathbf{x}^* = (1, \ldots, 1)^T$ with fitness $f(\mathbf{x}) = N$.

O. Kramer, *A Brief Introduction to Continuous Evolutionary Optimization*,
SpringerBriefs in Computational Intelligence,
DOI: 10.1007/978-3-319-03422-5_3, © The Author(s) 2014

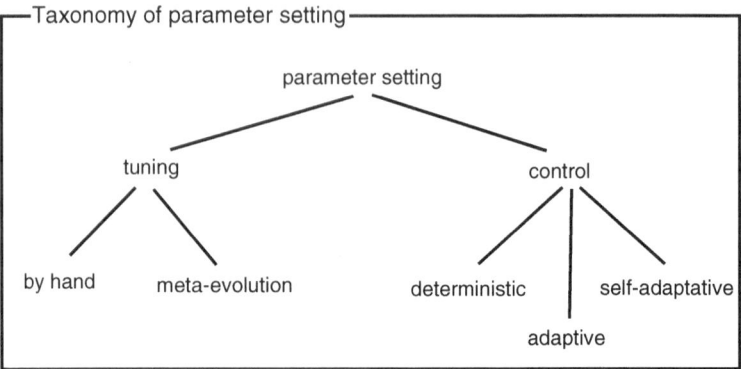

Fig. 3.1 Taxonomy of parameter setting of this work oriented to Eiben et al. [5] and complemented on the parameter tuning branch (cf. Kramer [6])

3.2 The (1+1)-EA

The $(1 + 1)$-EA works on bit string representations $\mathbf{x} = (x_1, \ldots, x_N)^T \in \{0, 1\}^N$ with only one individual, which is changed with bit-flip mutation. Bit-flip mutation means that each bit x_i of bit-string $\mathbf{x}$ is flipped with probability $\sigma = 1/N$. No recombination is employed, as no population is used. Furthermore, the selection operator can be reduced to a simple selection of the better one of two solutions. The pseudocode can be found in Algorithm 1. The number of fitness function calls of a $(1+1)$-EA complies with the number of generations.

Algorithm 1 Standard $(1 + 1)$-EA

1: choose $\mathbf{x} \in \{0, 1\}^N$ uniform at random
2: **repeat**
3:   produce $\mathbf{x}'$ by flipping each bit of $\mathbf{x}$ with probability $1/N$
4:   replace $\mathbf{x}$ with $\mathbf{x}'$ if $f(\mathbf{x}') \le f(\mathbf{x})$
5: **until** termination condition

For the $(1 + 1)$-EA, a runtime analysis on the simple *OneMax* problem demonstrates its properties. The runtime analysis is based on the method of fitness-based partitions, and shows that the $(1+1)$-EA's runtime is upper bounded by $O(N \log N)$ on *OneMax* [7].

Theorem 3.1 *The expected runtime of a $(1 + 1)$-EA on OneMax is $O(N \log N)$.*

The solution space $\{0, 1\}^N$ is divided into $N + 1$ sets $A_0, \ldots, A_N$. A partition A_i contains all solution with *OneMax*$(\mathbf{x}) = i$. If the currently best solution $\mathbf{x}$ belongs to A_{N-k}, still k 0-bits have to be flipped leading to improvements. The probability for another bit not to be flipped is $1 - \frac{1}{N}$, i.e., the probability that the ones are not flipped is $(1 - \frac{1}{N})^{(N-k)}$. Hence, the probability for success is at least $\frac{k}{N}(1 - \frac{1}{N})^{(N-k)} \geq \frac{k}{eN}$ for the next step. The expected runtime is upper bounded by eN/k. For different values of k, we get

$$\sum_{k=1}^{N} \frac{eN}{k} = eN \cdot \sum_{k=1}^{N} \frac{1}{k} = O(N \log N). \tag{3.2}$$

$\square$

In the remainder of this chapter, we will experimentally analyze and compare a selection of important parameter control and tuning techniques.

3.3 A Study on Mutation Rates

The question comes up, if our experiments can confirm the theoretical result, i.e., if the mutation rate $1/N$ leads to $N \log N$ generations in average. For this sake, we test the $(1 + 1)$-EA with various mutation rates on *OneMax* with various problem sizes. This extensive analysis is similar to tuning by hand, which is probably the most frequent parameter tuning method. Figure 3.2 shows the analysis with problem sizes $N = 10, 20$, and 30. The results show that the optimal mutation rate is close to $1/N$, which leads to the runtime of $O(N \log N)$. Our experiments confirm this result with the exception of a multiplicative constant, i.e., the runtime is about two times higher than $N \log N$. In the the following section, we employ evolutionary computation to search for optimal mutation rates, an approach called meta-evolution.

3.4 Meta-Evolution

Meta-evolution is a parameter tuning method that employs evolutionary computation to tune evolutionary parameters. The search for optimal parameters is treated as optimization problem. We employ a $(\mu + \lambda)$-ES [8] to tune the mutation rate of an inner $(1 + 1)$-EA. The $(\mu + \lambda)$-ES employs arithmetic recombination and isotropic

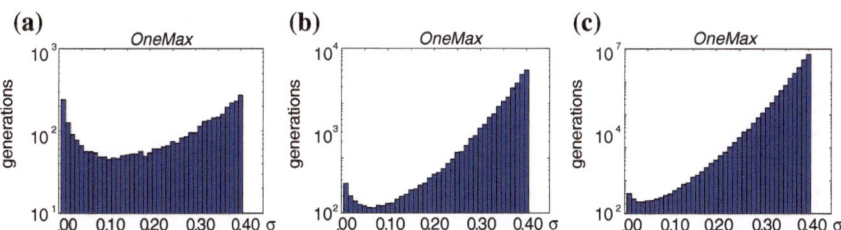

Fig. 3.2 Analysis of mutation strength σ for $(1 + 1)$-EA on *OneMax* for three problem sizes. **a** $(1 + 1)$-EA, $N = 10$, **b** $(1 + 1)$-EA, $N = 20$, **c** $(1 + 1)$-EA, $N = 30$

Table 3.1 Experimental results of meta-evolutionary approach of a $(10 + 100)$-EA tuning the mutation rates of a $(1 + 1)$-EA on *OneMax*

N	t	σ^*	Gen.
5	8.80	0.252987	37
10	31.84	0.134133	14
20	90.92	0.071522	42
30	170.60	0.055581	41

Gaussian mutation $\mathbf{x}' = \mathbf{x} + \mathcal{N}(0, \sigma)$ with a decreasing σ depending on generation t. Algorithm 2 shows the pseudocode of the meta-evolutionary approach.

Algorithm 2 Meta-$(1 + 1)$-EA

1: initialize mutation rates $\sigma_1, \ldots, \sigma_\mu \in \mathcal{P}, \tau$
2: **repeat**
3:    **for** $i = 1$ **to** λ **do**
4:        select ρ parents from $\mathcal{P}$
5:        create σ_i by recombination
6:        decrease τ
7:        mutate $\sigma_i = \sigma_i + \tau \cdot \mathcal{N}(0, 1)$
8:        run $(1 + 1)$-EA with σ_i
9:        add σ_i to $\mathcal{P}'$
10:   **end for**
11:   select μ parents from $\mathcal{P}' \to \mathcal{P}$
12: **until** termination condition

In our experimental analysis, we employ a $(10+100)$-ES optimizing the mutation rate of the underlying $(1 + 1)$-EA that solves problem *OneMax* for various problem sizes N. The ES starts with an initial mutation rate of $\tau = 0.2$. In each generation, τ is decreased deterministically by multiplication, i.e., $\tau = \tau \cdot 0.95$. The inner $(1+1)$-EA employs the evolved mutation rate σ of the upper ES and is run 25 times with this setting. The average number of generations until the optimum of *OneMax* is found employing the corresponding σ is the fitness $f(\sigma)$. The ES terminates after 50 generations. Table 3.1 shows the experimental results of the meta-evolutionary approach. The table shows the average number t of generations until the optimum has been found by the $(1+1)$-EA in the last generation of the ES, the evolved mutation rate σ^* and the number of generations, the ES needed to find σ^*. The achieved speed of convergence by the inner $(1 + 1)$-EA, e.g., 170.6 generations for $N = 30$ is a fast result.

3.5 Rechenberg's 1/5th Rule

An example for an adaptive control of endogenous strategy parameters is the 1/5th success rule for ES by Rechenberg [9]. The idea of Rechenberg's 1/5th rule is to increase the mutation rate, if the success probability is larger than 1/5th, and to decrease it, if the success probability is smaller. The success probability can be measured w.r.t. a fix number G of generations. If the number of successful generations, i.e., the offspring employs a better fitness than the parent, of a $(1 + 1)$-EA is g, then g/G is the success rate. If $g/G > 1/5$, σ is increased by $\sigma = \sigma \cdot \tau$ with $\tau > 1$, otherwise, it is decreased by $\sigma = \sigma/\tau$. Algorithm 3 shows the pseudocode of the $(1 + 1)$-EA with Rechenberg's 1/5th rule. The objective is to stay in the so called *evolution window* guaranteeing nearly optimal progress.

Algorithm 3 $(1 + 1)$-EA with Rechenberg's 1/5th rule

1: choose $\mathbf{x} \in \{0, 1\}^N$ uniform at random
2: **repeat**
3:   **for** $i = 1$ to G **do**
4:     produce $\mathbf{x}'$ by flipping each bit of $\mathbf{x}$ with probability σ
5:     replace $\mathbf{x}$ with $\mathbf{x}'$ if $f(\mathbf{x}') \le f(\mathbf{x})$ and set $g+ = 1$
6:   **end for**
7:   **if** $g/G > 1/5$ **then**
8:     $\sigma = \sigma \cdot \tau$
9:   **else**
10:     $\sigma = \sigma/\tau$
11:   **end if**
12: **until** termination condition

Figure 3.3 shows the corresponding experimental results for various values of τ and $N = 10, 20$, and 30. The results show that Rechenberg's rule is able to automatically tune the mutation rate and reach almost as good results as the runs with tuned settings. We can observe that smaller settings for τ, i.e., settings close to 1.0 achieve better results than larger settings in all cases. Further experiments have shown that settings over $\tau > 10.0$ lead to very long runtimes (larger than 10^5 generations). In such cases, σ cannot be fine-tuned to allow a fast approximation of the optimum.

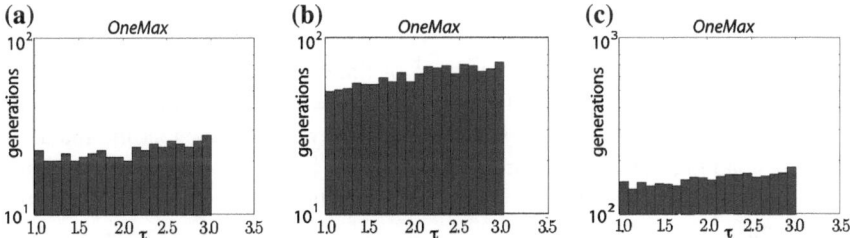

Fig. 3.3 Experimental results of parameter control with Rechenberg's 1/5th rule. **a** Rechenberg, $N = 5$, **b** Rechenberg, $N = 10$, **c** Rechenberg, $N = 20$

3.6 Self-Adaptation

Self-adaptation is an automatic evolutionary mutation rate control. It was originally introduced by Rechenberg and Schwefel [10] for ES, later independently in the United States by Fogel [11] for evolutionary programming. The most successful self-adaptively controlled parameters are mutation parameters. This is a consequence of the direct influence of the mutation operator on the exploration behavior of the optimization algorithm: Large mutation strengths cause large changes of solutions, decreasing mutation strengths allow an approximation of the optimum, in particular in continuous solution spaces.

The mutation rate σ is added to each individual $\mathbf{x}$ and is at the same time subject to recombination, mutation and selection. For a $(1 + 1)$-EA, self-adaptation means that the mutation rate σ is mutated with log-normal mutation

$$\sigma' = \sigma \cdot e^{\tau \mathcal{N}(0,1)} \tag{3.3}$$

with a control parameter τ. Afterwards, the mutation operator is applied. Appropriate mutation rates are inherited and employed in the following generation. The log-normal mutation allows an evolutionary search in the space of strategy parameters. It allows the mutation rates to scale in a logarithmic kind of way from values close to zero to infinity. Algorithm 4 shows the pseudocode of the SA-$(1 + 1)$-EA with σ-self-adaptation.

Algorithm 4 SA-$(1 + 1)$-EA

1: choose $\mathbf{x} \in \{0, 1\}^N$ uniform at random
2: choose $\sigma \in \{0, 1\}$ at random
3: **repeat**
4:    produce $\sigma' = \sigma \cdot e^{\tau \mathcal{N}(0,1)}$
5:    produce $\mathbf{x}'$ by flipping each bit of $\mathbf{x}$ with probability σ'
6:    replace $\mathbf{x}$ with $\mathbf{x}'$ and σ with σ', if $f(\mathbf{x}') \leq f(\mathbf{x})$
7: **until** termination condition

Figure 3.4 shows typical developments[1] of fitness $f(\mathbf{x})$ and mutation rate σ of the SA-$(1 + 1)$-EA on $N = 10, 50$, and 100 for $\tau = 0.1$. Due to the plus selection scheme, the fitness is decreasing step by step. The results show that the mutation rate σ is adapting during the search. In particular, in the last phase of the search for $N = 100$, σ is fast adapting to the search conditions and accelerates the search significantly.

Table 3.2 shows the experimental results of the SA-$(1+1)$-EA with various settings for τ on *OneMax* with problem sizes $N = 10, 20, 30, 50$, and 100. The results show that the control parameter, i.e., the mutation rate τ of the mutation rate σ, has a significant impact on the success of the SA-$(1 + 1)$-EA. Both other setting, i.e.,

---

[1] employing a logarithmic scale

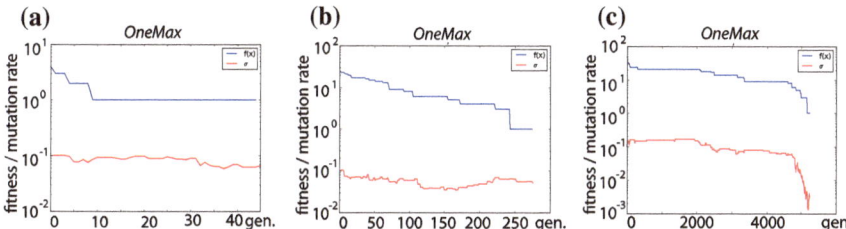

Fig. 3.4 SA-(1 + 1)-EA on *OneMax* with N = 10,50, and 100. **a** SA, N = 10, **b** SA, N = 50, **c** SA, N = 100

Table 3.2 Number of generations the SA-(1 + 1)-EA needs to reach the optimum

N	10	20	30	50	100
$\tau = 0.01$	48.3 ± 29.03	162.0 ± 83.1	359.0 ± 175.0	$2.4e3 \pm 552.8$	$> 10^5$
$\tau = 0.1$	46.1 ± 36.3	142.9 ± 47.1	274.0 ± 97.4	$1.0e3 \pm 770.7$	$3.6e3 \pm 3.3e3$
$\tau = 1.0$	$2.7e3 \pm 4.9e3$	$5.0e3 \pm 1.2e4$	$8.9e3 \pm 9.5e3$	$1.9e4 \pm 1.4e4$	$> 10^5$

$\tau = 0.01$ and $\tau = 1.0$ lead to worse results. In particular on the large problem instance with $N = 100$, both settings fail and lead to long optimization runs.

3.7 Conclusions

The success of evolutionary algorithms depends on the choice of appropriate parameter settings, in particular mutation rates. Although a lot of studies are known in literature, only few compare different parameter control techniques employing the same algorithmic settings on the same problems. But only such a comparison allows insights into the underlying mechanisms and common principles. The analysis has shown that optimally tuned mutation rates can automatically be found with meta-evolution. The effort spent into the search is comparatively high, but the final result is competitive or better than the control techniques. But more flexible and still powerful is the adaptive mutation rate control with Rechenberg's rule. Self-adaptation turns out to be the most flexible control technique with its automatic mutation rate control. Although self-adaptation depends on the control parameter τ, it is quite robust w.r.t. the problem size. It became famous in ES for continuous optimization and also has shown the best results in our parameter control study. As future work, we plan to extend our analysis to further EA variants, parameter control techniques, and problem types.

References

1. K.A.D. Jong, An analysis of the behavior of a class of genetic adaptive systems. Ph.D. thesis, University of Michigan, 1975
2. J.D. Schaffer, R. Caruana, L.J. Eshelman, R. Das, A study of control parameters affecting online performance of genetic algorithms for function optimization, in *Proceedings of the 3rd International Conference on Genetic Algorithms (ICGA)*, pp. 51–60, 1989
3. J. Grefenstette, Optimization of control parameters for genetic algorithms. IEEE Trans. Syst. Man Cybern. **16**(1), 122–128 (1986)
4. H. Mühlenbein, How genetic algorithms really work: Mutation and hillclimbing, in *Proceedings of the 2nd Conference on Parallel Problem Solving from Nature (PPSN)*, pp. 15–26, 1992
5. A.E. Eiben, R. Hinterding, Z. Michalewicz, Parameter control in evolutionary algorithms. IEEE Trans. Evol. Comput. **3**(2), 124–141 (1999)
6. O. Kramer, *Self-Adaptive Heuristics for Evolutionary Computation, Studies in Computational Intelligence* (Springer, Heidelberg, 2008)
7. S. Droste, T. Jansen, I. Wegener, On the analysis of the (1+1) evolutionary algorithm. Theoret. Comput. Sci. **276**(1–2), 51–81 (2002)
8. H.-G. Beyer, H.-P. Schwefel, Evolution strategies—A comprehensive introduction. Nat. Comput. **1**, 3–52 (2002)
9. I. Rechenberg, *Evolutionsstrategie: Optimierung Technischer Systeme nach Prinzipien der Biologischen Evolution* (Frommann-Holzboog, Stuttgart, 1973)
10. H.-P. Schwefel, *Adaptive Mechanismen in der biologischen Evolution und ihr Einfluss auf die Evolutionsgeschwindigkeit* (Interner Bericht der Arbeitsgruppe Bionik und Evolutionstechnik am Institut für Mess- und Regelungstechnik, TU Berlin, 1974)
11. D.B. Fogel, L.J. Fogel, J.W. Atma, Meta-evolutionary programming, in *Proceedings of 25th Asilomar Conference on Signals, Systems and Computers*, pp. 540–545, 1991

Part II
Advanced Optimization

Chapter 4
Constraints

4.1 Introduction

Constraints can make a hard optimization problem even harder. They restrict the solution space to a feasible subspace. In practice, constraints are typically not considered available in their explicit formal form, but are assumed to be black boxes: a vector $\mathbf{x} \in \mathbb{R}^N$ fed to the black box just returns a numerical or boolean value stating the constraint violation. In this chapter, we concentrate on constraints that are not given explicitly, a case often encountered in complex simulation models. The constrained real-valued optimization problem is to find a solution $\mathbf{x} \in \mathbb{R}^N$ in the N-dimensional solution space $\mathbb{R}^N$ that minimizes the objective function, i.e.:

$$
\begin{aligned}
&\text{minimize} \quad f(\mathbf{x}), \qquad \mathbf{x} \in \mathbb{R}^N \text{ subject to} \\
&\text{inequalities } g_i(\mathbf{x}) \le 0, \ i = 1, \dots, n_1, \text{ and} \\
&\text{equalities} \quad h_j(\mathbf{x}) = 0, \ j = 1, \dots, n_2 .
\end{aligned} \tag{4.1}
$$

We measure the constraint violation with

$$
G(\mathbf{x}) = \sum_{i=1}^{n_1} \max(0, g_i(\mathbf{x})) + \sum_{j=1}^{n_2} |h_j(\mathbf{x})|. \tag{4.2}
$$

Coello [1] and Kramer [2] are good starting points for literature surveys on constraint handling methods for EAs. Most methods fall into the category of penalty functions, e.g., the penalty function proposed by Kuri and Quezada et al. [3] that allows the search process to discover the whole solution space, penalizing the infeasible part. Other approaches are based on decoders, see Michalewizc [4].

Penalty functions deteriorate the fitness of infeasible solutions by taking the number of unfulfilled constraints or the distance to feasibility into account [3, 5–7]. In this chapter, we introduce an adaptive penalty function that allows to handle difficult constrained problems. The adaptation process weakens and strengthens the penalty based on the number of feasible and infeasible solutions in one generation.

O. Kramer, *A Brief Introduction to Continuous Evolutionary Optimization*,
SpringerBriefs in Computational Intelligence,
DOI: 10.1007/978-3-319-03422-5_4, © The Author(s) 2014

An early penalty function is the sequential unconstrained minimization technique by Fiacco and McCormick [5]. It is based on minimizing a sequences of constrained problems with stepwise intensified penalty factors. Similar approaches employ static penalty factors, e.g., the one proposed by Homaifar et al. [6] or intensify the penalty w.r.t. the number of satisfied constraints, e.g., Kuri and Quezada's approach [3]. Dynamic penalty functions intensify the penalty depending on the number of generations, e.g., the penalty function by Joines and Houck [7]

$$\tilde{f}(\mathbf{x}) = f(\mathbf{x}) + \gamma \cdot G(\mathbf{x}) \tag{4.3}$$

with $\gamma = (C \cdot t)^{\alpha}$ at generation t and user-defined parameters C and α. Typical settings are $C = 0.5$ and $\alpha = 1$. Penalties can also be adapted according to an external cooling scheme [7]. In the segregated genetic algorithm by Le Riche et al. [8], two penalty functions, a weak and an intense one, are computed in order to surround the optimum.

Adaptive penalties employ features collected during the evolutionary search. The adaptive penalty approach of this work is based on the number of feasible solutions. An early adaptive penalty function is the approach by Bean and Hadj-Alouane [9] that adapts the penalty factor as follows

$$\gamma' = \begin{cases} (1/\tau_1) \cdot \gamma, & \text{if } G(\mathbf{x}_j^*) = 0 \text{ for all } t - k + 1 \leq j \leq t \\ \tau_2 \cdot \gamma, & \text{if } G(\mathbf{x}_j^*) > 0 \text{ for all } t - k + 1 \leq j \leq t \\ \gamma, & \text{else} \end{cases} \tag{4.4}$$

with parameters $\tau_1, \tau_2 > 1$ and $\mathbf{x}_j^*$ being the best solution w.r.t. $\tilde{f}(\cdot)$ in the last k generations, i.e., generations $j = t - k + 1, \ldots, t$.

4.2 Adaptive Penalty Function

The adaptive penalty function that is core of our constraint handling approach is introduced in this section. It will be experimentally analyzed before it is integrated into a meta-model learning process. The penalty function is oriented to Rechenberg's 1/5th success rule, cf. Chap. 4. Often, the optimal solution of a constrained problem lies in the vicinity of the feasible solution space. To let the search explore this region, penalty functions have to balance the penalty factors. This can be accomplished with a Rechenberg-like rule. If less than 1/5th of the population is feasible, the penalty factor γ is increased to move the population into the feasible region

$$\gamma = \gamma \cdot \tau \tag{4.5}$$

with $\tau > 1$. Otherwise, i.e., if more than 1/5th of the population of candidate solutions is feasible, the penalty is weakened

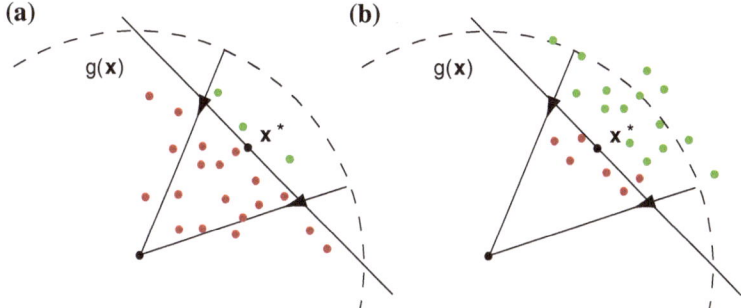

Fig. 4.1 Penalties are **a** increased, if less than 1/5th of the population is feasible and **b** decreased, if more than 1/5th is feasible

$$\gamma = \gamma/\tau \qquad (4.6)$$

to allow to move into the infeasible part of the solution space. The success rate of 1/5th allows the fastest progress towards the optimal solution. Figure 4.1 illustrates the situation. In Fig. 4.1a, less than 1/5th of the population is feasible. The penalty should be increased to move the search into the feasible solution space. Figure 4.1b shows the situation that more than 1/5th of the population is feasible. To move the search into the infeasible region, the penalty factor should be decreased.

In our experimental analysis, we focus on the difficult *Tangent* problem (f_{TR}), cf. Appendix. The *Tangent* problem is difficult to solve, as the linear constraint is a tangent to the contours of the *Sphere* function [10] (cf. Fig. 4.2). The closer the search comes to the constraint, the more parallel are the contour lines of the *Sphere* function to the tangent and the more does the gradient point into the direction of the infeasible solution space.

4.3 Experimental Analysis

In this section, we perform an experimental analysis of the proposed Rechenberg penalty function based on an (μ, λ)-ES with isotropic σ-self-adaptive step sizes. The adaptive penalty function depends on the magnitude of the penalty factor change. In the first part of our experimental analysis, we analyze the influence of τ systematically. We employ the following experimental settings:

- (μ, λ)-ES,
- isotropic σ-self-adaptation,
- initial candidate solution interval $\mathbf{x} \in [-10^4, 10^4]^N$,
- initial candidate step size $\sigma = 10.0$,
- initial penalty factor $\gamma = 2^4$,
- termination, if the best feasible solution reaches the optimum with accuracy $\theta = 10^{-10}$.

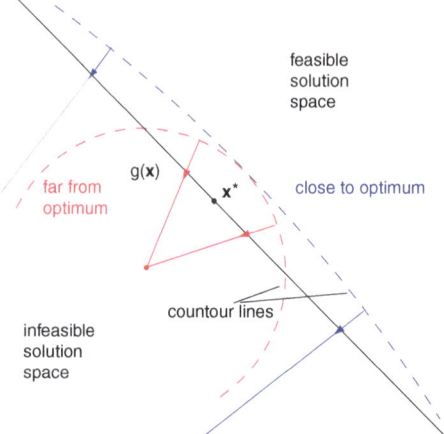

Fig. 4.2 Contour lines at two situations near the optimum of f_{TR} with $N = 2$. If the search is comparatively far away from the optimum (*red*), the gradient, i.e., the direction to the unconstrained optimum is approximately pointing into the direction of the real optimum $\mathbf{x}^*$, while the gradient is almost orthogonally pointing to the constraint boundary, if the search takes place at the constraint boundary (*blue*), closely to the optimum

Table 4.1 shows the analysis of the number of fitness function evaluations w.r.t. different problem space dimensions on f_{TR}. We can observe that the simple evolution strategy with isotropic Gaussian mutation is able to approximate the optimum of f_{TR}, even with high dimensions. The adaptive Rechenberg-like penalty factor adaptation mechanism is obviously able to balance the search at the boundary of the infeasible solution space. However, it turns out that the choice of appropriate values for τ has an important part to play for this success. Only settings very close to $\tau = 2.0$ allow the approximation of the optimal solution.

We tested the same experimental setting on problem $f_{2.40}$ and discovered that different values for τ are necessary. Table 4.2 shows the experimental results. The most important observation is that the adaptive penalty function allows the evolution strategy to approximate the optimum. It turns out that the best setting is $\tau = 30$. We analyze the evolutionary runs with adaptive penalty function and different parameterizations on the *Tangent* problem with varying dimensions. For this sake, the same settings are used like in the pervious section. Figure 4.3 shows the develop-

Table 4.1 Analysis of evolution strategy with dynamic penalty function on the *Tangent* problem with increasing dimensions, the mean values of fitness function and constraint function evaluations of 100 runs of the best fitness achieved are shown

N	2		10		50	
	ffe	cfe	ffe	cfe	ffe	cfe
f_{TR}	84,816	85,870	407,002	412,068	1,408,138	1,425,669

Table 4.2 Analysis of the influence of parameter τ on mean number of fitness and constraint function evaluation of 100 runs on problem $f_{2.40}$

τ	20.0		30.0		40.0	
	ffe	cfe	ffe	cfe	ffe	cfe
$f_{2.40}$	5,000,394	5,061,894	4,397,045	4,451,343	7,004,593	7,090,978

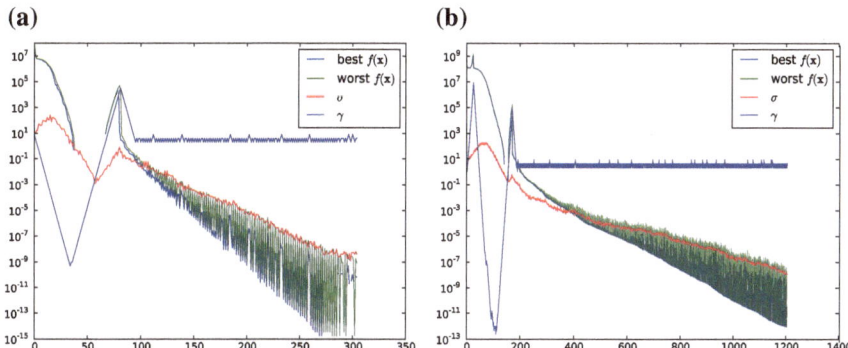

Fig. 4.3 Analysis of the adaptive penalty function on the *Tangent* problem w.r.t. two problem dimensions, i.e., $N = 10$ and $N = 50$. The plots show typical runs of the development of the best and worst fitness, step sizes, and γ on a log-scale w.r.t. the generation number. **a** f_{TR}, $N = 10$; **b** f_{TR}, $N = 50$

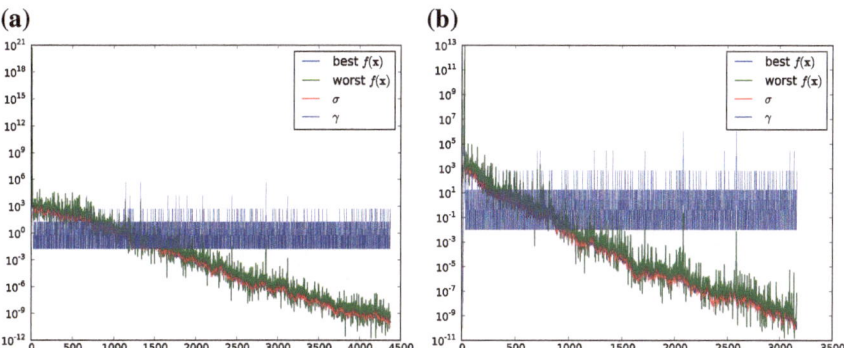

Fig. 4.4 Analysis of the adaptive penalty function on problem $f_{2.40}$ with settings $\tau = 30$ and $\tau = 40$. **a** $f_{2.40}$, $\tau = 30.0$ **b** $f_{2.40}$, $\tau = 40.0$

ment on the f_{TR} problem, while Fig. 4.4 shows the corresponding results on problem $f_{2.40}$ for two settings of τ.

Remarkable is the first part of the search on the tangent problem. Step sizes are increasing, i.e., the self-adaptive step size mechanism allows bigger steps. The penalty factor is deceased, and the search moves into the infeasible region. Obviously, too few solutions are in the infeasible solution space. Then, the steps are decreasing, while the penalty factor is increased again to move the search into the feasible region. Afterwards, the search performs a perfect log-linear approximation of the optimum.

4.4 Meta-Modeling

In this section, we decrease the number of fitness and constraint function evaluations with two meta-models, i.e., a fitness function and a constraint meta-model. Optimization procedures in design processes may require a large number of function (response) evaluations. Often, due to complex relationships between models, analytic relationships cannot be determined, leading to the black box optimization scenario. A significant reduction of computational effort (e.g., spent on complex simulation models) can be achieved employing meta-modeling techniques. Meta-models are machine learning models used as surrogates of the real simulation model. Based on past response evaluations, a statistical model is built that serves as basis for response evaluation estimates. This idea is related to the standard supervised learning scenario. A meta-model of a constrained test function will be used to handle constraints more effectively. To learn an effective meta-model, strategies to generate sampling points are necessary. For this sake, various methods have been presented, e.g., methods from design of experiments that try to maximize the information between input and functional response, or active learning methods. A very successful meta-model that has often been applied in computational engineering is Kriging [11], which belongs to the class of linear least squares estimation algorithms.

The meta-model we employ is KNN for fitness and constraint function meta-modeling. KNN is a supervised machine learning algorithm that predicts the label of unknown patterns based on the pattern-label pairs in a training set corresponding to a regression function $\mathbf{f} : \mathbb{R}^q \to \mathbb{R}^d$.

$$\mathbf{f}_{KNN}(\mathbf{x}') = \frac{1}{K} \sum_{i \in \mathfrak{N}_K(\mathbf{x}')} \mathbf{y}_i \qquad (4.7)$$

with set $\mathfrak{N}_K(\mathbf{x}')$ containing the indices of the K-nearest neighbors of $\mathbf{x}'$. We employ two KNN regression meta-models, one for the fitness function and one for the constraint violation.

They are integrated into the optimization process as follows. Both employ an archive of the last $n = 100$ fitness function and constraint violation values. Let $\mathbf{x}_1, \ldots, \mathbf{x}_t$ be the sequence of solutions, which have been evaluated on $f(\cdot)$ and $G(\cdot)$. For a new function evaluation $f(\mathbf{x}_{t+1})$, the archive A_f of fitness function evaluations is updated as follows

$$A_f = \{(\mathbf{x}_{t-n+2}, f(\mathbf{x}_{t-n+2})), \ldots, (\mathbf{x}_{t+1}, f(\mathbf{x}_{t+1}))\} \qquad (4.8)$$

Table 4.3 Analysis of ES with fitness function and constraint function meta-models in combination with the adaptive penalty function in 100 runs on problem f_{TR} with three dimensions and three settings for β

mm	no		20	
N	ffe	cfe	ffe	cfe
2	99,925	101,197	64,562	65,580
5	160,974	199,232	189,830	192,914
10	372,294	377,013	321,438	326,713
	30		50	
2	61,869	62,934	64,124	67,345
5	152,783	155,520	170,264	173,817
10	405,972	413,281	305,514	311,969

and the corresponding update of the archive of constraint evaluations

$$A_c = \{(\mathbf{x}_{t-n+2}, G(\mathbf{x}_{t-n+2})), \dots, (\mathbf{x}_{t+1}, G(\mathbf{x}_{t+1}))\}, \tag{4.9}$$

respectively. Every 100 function evaluations, β evaluations are evaluated on the meta-models. We test various settings for β in the next section.

In this section, we analyze the evolution strategy with adaptive penalty function and two meta-models. Table 4.3 shows the corresponding experimental results (median of fitness and constraint function evaluations) on the *Tangent* problem with three different dimensions, i.e., $N = 2, 5, 10$ and various settings for β. We can make the following observations. The meta-models decrease the number of fitness and constraint function evaluations. The advantage of the employment of meta-models is higher for small N than for larger N. The savings of function evaluations are a good reason to employ meta-models in constrained solution spaces.

4.5 Conclusions

The difficult *Tangent* problem can be solved with the proposed simple adaptive Rechenberg-like penalty function. Penalties are increased, if less than 1/5th of the population is feasible, while they are decreased otherwise. An experimental analysis has proven the capabilities of the approach. The adaptive penalty function allows the approximation even for higher problem dimensions, which has been shown to be very difficult in previous work. Also on the difficult problem $f_{2.40}$, the penalty function has shown comparatively good results. The two meta-models accelerate the search in terms of fitness and constraint function evaluations. Our future work will concentrate on multi-objective and dynamic constrained problems, where the adaptive penalty function is potentially a promising approach.

References

1. C.A. Coello Coello, Theoretical and numerical constraint handling techniques used with evolutionary algorithms: a survey of the state of the art. Comput. Methods Appl. Mech. Eng. **191**(11–12), 1245–1287 (2002)
2. O. Kramer, *Self-Adaptive Heuristics for Evolutionary Computation, Studies in Computational Intelligence* (Springer, Heidelberg, 2008)
3. A. Kuri-Morales, C.V. Quezada, A universal eclectic genetic algorithm for constrained optimization, in *Proceedings 6th European Congress on Intelligent Techniques and Soft Computing (EUFIT)*, (Aachen, Verlag Mainz, 1998), pp. 518–522
4. Z. Michalewicz, D.B. Fogel, *How to Solve It: Modern Heuristics* (Springer, Berlin, 2000)
5. A. Fiacco, G. McCormick, The sequential unconstrained minimization technique for nonlinear programming—a primal-dual method. Mgmt. Sci. **10**, 360–366 (1964)
6. A. Homaifar, S.H.Y. Lai, X. Qi, Constrained optimization via genetic algorithms. Simulation **62**(4), 242–254 (1994)
7. J. Joines, C. Houck, On the use of non-stationary penalty functions to solve nonlinear constrained optimization problems with GAs, in *Proceedings of the 1st IEEE Conference on Evolutionary Computation* (IEEE Press, Orlando, 1994), pp. 579–584
8. R.G.L. Riche, C. Knopf-Lenoir, R.T. Haftka, A segregated genetic algorithm for constrained structural optimization, in *Proceedings of the 6th International Conference on Genetic Algorithms (ICGA)* (University of Pittsburgh, Morgan Kaufmann Publishers, San Francisco, 1995), pp. 558–565
9. J.C. Bean, A.B. Hadj-Alouane, A dual genetic algorithm for bounded integer programs. Technical report, University of Michigan, 1992
10. O. Kramer, Premature convergence in constrained continuous search spaces, in *Proceedings of the 10th Conference on Parallel Problem Solving from Nature (PPSN)*, LNCS (Springer, Berlin, 2008), pp. 62–71
11. D.G. Krige, A statistical approach to some mine valuation and allied problems on the Witwatersrand. Master's thesis. University of the Witwatersrand, South Africa, 1951

Chapter 5
Iterated Local Search

5.1 Introduction

Hybridization has developed to an effective strategy in algorithm design. Hybrid algorithms can become more efficient and more effective than their *native* counterparts. This observation holds true for many problem classes, in particular in optimization, where hybrid techniques of meta-heuristics and local search are often called hybrid meta-heuristics. In this chapter, we show how Powell's conjugate gradient search, which is a fast and powerful black box optimization strategy for convex problems, can be integrated into an ES [1]. Further, we show how to employ a specialized step size adaptation technique that allows to guide the optimization process and to escape from local optima that Powell's method may successively find.

5.2 Iterated Local Search

Iterated local search (ILS) is based on a simple but successful idea. Instead of repeating local search and starting from initial solutions like restart approaches do, ILS begins with a solution $\mathbf{x}$ and successively applies local search and perturbation of the local optimal solution $\hat{\mathbf{x}}$. This procedure is repeated iteratively until a termination condition is fulfilled. Algorithm 1 shows the pseudocode of the ILS approach. Initial solutions should use as much information as possible to be a good starting point for local search. Most local search operators are deterministic. Consequently, the perturbation mechanism should introduce non-deterministic components to explore the solution space. The perturbation mechanism performs global random search in the space of local optima that are approximated by the local search method. Blum and Roli [2] point out that the balance of the perturbation mechanism is quite important. Perturbation must be strong enough to allow the escape from basins of attraction, but weak enough to exploit knowledge from previous iterations. Otherwise, the ILS will

O. Kramer, *A Brief Introduction to Continuous Evolutionary Optimization*,
SpringerBriefs in Computational Intelligence,
DOI: 10.1007/978-3-319-03422-5_5, © The Author(s) 2014

become a simple restart strategy. The acceptance criterion of Line 6 may vary from *always accept* to *only accept in case of improvement*. Approaches like simulated annealing may be adopted.

Algorithm 1 Iterated Local Search

1: initialize solution **x**
2: produce **x̂** with local search
3: **repeat**
4:    perturbation of **x**
5:    produce **x̂** with local search
6:    apply acceptance criterion
7: **until** termination condition

There are many examples in literature for the successful application of ILS variants on combinatorial optimization problems. A survey of ILS techniques has been presented by Lourenco et al. [3]. The authors also provide a comprehensive introduction [4] to ILS. A famous combinatorial instance, many ILS methods have been developed for, is the traveling salesperson problem. Stützle and Hoos [5] introduced an approach that combines restarts with a specific acceptance criterion to maintain diversity for the TSP, while Katayama and Narihisa [6] use a perturbation mechanism that combines the heuristic 4-opt with a greedy method. Stützle [7] uses an ILS hybrid to solve the quadratic assignment problem. The technique is enhanced by acceptance criteria that allow moves to worse local optima. Furthermore, population-based extensions are introduced. Duarte et al. [8] introduce an ILS heuristic for the problem of assigning referees to scheduled games in sports based on greedy search. Our perturbation mechanism is related to their approach. Preliminary work on the adaptation of the perturbation algorithm has been applied by Mladenovic et al. [9] for variable neighborhood search and tabu search by Glover et al. [10].

5.3 Powell's Conjugate Gradient Method

The hybrid ILS variant introduced in this chapter is based on Powell's optimization method. Preliminary experiments revealed the efficiency of Powell's method in comparison to continuous evolutionary search methods. However, in the experimental section, we will observe that Powell's method can get stuck in local optima in multimodal solution spaces. An idea similar to the hybridization of local search has been presented by *Griewank* [11], who combines a gradient descent method with a deterministic perturbation term.

Powell's method belongs to the class of direct search methods, i.e., no first or second order derivatives are required. It is based on conjugate directions and is similar to line search. The idea of line search is to start from search point $\mathbf{x} \in \mathbb{R}^N$ along a direction $\mathbf{d} \in \mathbb{R}^N$, so that $f(\mathbf{x} + \lambda_t \mathbf{d})$ is minimized for a $\lambda_t \in \mathbb{R}^+$. Powell's method [12, 13] adapts the directions according to a gradient-like information from the search.

Algorithm 2 Powell's Method

1: **repeat**
2:   **for** $t = 1$ to N **do**
3:     find λ_t that minimizes $f(\mathbf{x}_{t-1} + \lambda_t \mathbf{d}_t)$
4:     set $\mathbf{x}_t = \mathbf{x}_{t-1} + \lambda_t \mathbf{d}_t$
5:     **for** $j = 1$ to $N - 1$ **do**
6:       update vectors $\mathbf{d}_j = \mathbf{d}_{j+1}$
7:     **end for**
8:     set $\mathbf{d}_N = \mathbf{x}_N - \mathbf{x}_0$
9:     find λ_N that minimizes $f(\mathbf{x}_N + \lambda_N \mathbf{d}_N)$
10:     set $\mathbf{x}_0 = \mathbf{x}_0 + \lambda_N \mathbf{d}_N$
11:   **end for**
12: **until** termination condition

It is based on the assumption of a quadratic convex objective function $f(\mathbf{x})$

$$f(\mathbf{x}) = \frac{1}{2}\mathbf{x}^T \mathbf{H} \mathbf{x} + \mathbf{b}^T \mathbf{x} + c. \tag{5.1}$$

with Hessian matrix $\mathbf{H}$. Two directions $\mathbf{d}_i, \mathbf{d}_j \in \mathbb{R}^N$, $i \neq j$ are mutually conjugate, if

$$\mathbf{d}_i^T \mathbf{H} \mathbf{d}_j = 0 \tag{5.2}$$

holds with mutual conjugate directions that constitute a basis of the solution space $\mathbb{R}^N$. Let $\mathbf{x}_0$ be the initial guess of a minimum of function f. In iteration t, we require an estimation of the gradient $\mathbf{g}_t = \mathbf{g}(\mathbf{x}_t)$. Let $t = 1$ and let $\mathbf{d}_t = -\mathbf{g}_t$ be the steepest descent direction. For $t > 1$, Powell applies the equation

$$\mathbf{d}_t = -\mathbf{g}_t + \beta_t \mathbf{d}_{t-1}, \tag{5.3}$$

with the Euclidean vector norms

$$\beta_t = \frac{\|\mathbf{g}_t\|^2}{\|\mathbf{g}_{t-1}\|^2}. \tag{5.4}$$

The main idea of the conjugate direction method is to search for the minimal value of $f(\mathbf{x})$ along direction $\mathbf{d}_t$ to obtain the next solution $\mathbf{x}_{t+1}$, i.e., to find the λ that minimizes

$$f(\mathbf{x}_t + \lambda_t \mathbf{d}_t). \tag{5.5}$$

For a minimizing λ_t, set

$$\mathbf{x}_{t+1} = \mathbf{x}_t + \lambda_t \mathbf{d}_t. \tag{5.6}$$

Algorithm 2 shows the pseudocode of the conjugate gradient method that is the basis of Powell's strategy. In our implementation, the search for λ_t is implemented with line search. For a more detailed introduction, we refer to the depiction of Powell [12] and Schwefel [14].

At first, we analyze Powell's method on the optimization test suite (cf. Appendix A). Solutions are randomly initialized in the interval $[-100, 100]^N$. Each experiment is repeated 30 times. Powell's method terminates, if the improvement from one to the next iteration is smaller than $\phi = 10^{-10}$ with comma selection, or if the optimum is found with accuracy $f_{stop} = 10^{-10}$. As Powell's method is a convex optimization technique, we expect that only the unimodal problems can be solved. Table 5.1 confirms these expectations. On unimodal functions, Powell's method is exceedingly fast. On the *Sphere* problem with $N = 10$, a budget of only 101.7 fitness function evaluations in mean is sufficient to approximate the optimum. These fast approximation capabilities can also be observed on problems *Doublesum* and *Rosenbrock*, also for higher dimensions, i.e. $N = 30$.

The results also show that Powell's method is not able to approximate the optima of the multimodal function *Rastrigin*. On the *easier* multimodal function *Griewank*, the random initializations allow to find the optimum in some of the 30 runs. The fast convergence behavior on convex function parts motivates to perform local search as operator in a global evolutionary optimization framework. It is the basis of the Powell ES that we will analyze in the following.

Table 5.1 Experimental comparison of Powell's method on the test problems with $N = 10$ and $N = 30$ dimensions

	Best	Median	Worst	Mean	Dev	#
$N = 10$						
f_{Sp}	100	102	102	101.7	0.67	30
f_{Dou}	91	92	92	91.8	0.42	30
f_{Ros}	2,947	4,617	12,470	5,941.87	3,353.14	24
f_{Ras}	–	–	–	–	–	0
f_{Gri}	329	329	329	329	0	1
f_{Kur}	–	–	–	–	–	0
$N = 30$						
f_{Sp}	299	302	302	301.3	1.05	30
f_{Dou}	291	291.5	292	291.5	0.52	30
f_{Ros}	14,888	33,315	59,193	36,455.85	16,789.41	21
f_{Ras}	–	–	–	–	–	0
f_{Gri}	904	997	1,001	967.33	54.88	3
f_{Kur}	–	–	–	–	–	0

Best, median, worst, mean, and *dev* provide statistical information about the number of fitness function evaluations of 30 runs until the difference between the fitness of the best solution and the optimum is smaller than $f_{stop} = 10^{-10}$. Parameter # states the number of runs that reached the optimum.

5.4 Powell Evolution Strategy

The Powell ES [1] presented in this section is based on four key concepts, each focusing on typical problems in real-valued solution spaces. Powell's method is a fast direct optimization method, in particular appropriate for unimodal fitness landscapes. It is integrated into the optimization process using ILS, in order to prevent Powell's method from getting stuck in local optima. ILS approach is based on the successive repetition of Powell's conjugate gradient method as local search technique and a perturbation mechanism. A population of candidate solutions is employed for exploration similar to evolution strategies. The strength of the ILS perturbation is controlled by means of an adaptive control mechanism. In case of stagnation, the mutation strength is *increased*, in order to leave local optima, and *decreased* otherwise.

Algorithm 3 shows the pseudocode of the Powell ES. At the beginning, μ solutions $\mathbf{x}_1, \ldots, \mathbf{x}_\mu \in \mathbb{R}^N$ are randomly initialized and optimized with the strategy of Powell. In an iterative loop λ, offspring solutions $\mathbf{x}_1, \ldots, \mathbf{x}_\lambda$ are produced by means of Gaussian mutations with the global mutation strength σ by

$$\mathbf{x}'_j = \mathbf{x}_j + \mathbf{z}_j, \tag{5.7}$$

with

$$\mathbf{z}_j \sim (\sigma_1 \mathcal{N}(0, 1), \ldots, \sigma_N \mathcal{N}(0, 1))^T. \tag{5.8}$$

Afterwards, each solution $\mathbf{x}'_j$ is locally optimized with the strategy of Powell, leading to $\hat{\mathbf{x}}'_j$ for $j = 1, \ldots, \lambda$. After λ solutions have been produced this way, the μ-best are selected according to their fitness with comma selection. Then, we apply global recombination, i.e., the arithmetic mean $\langle \hat{\mathbf{x}}_t \rangle$ at generation t of all selected solutions $\hat{\mathbf{x}}_1, \ldots, \hat{\mathbf{x}}_\mu$ is computed. The fitness of this arithmetic mean is evaluated and compared to the fitness of the arithmetic mean of the last generation $t-1$. If the search stagnates, i.e., if the condition

$$|f(\langle \hat{\mathbf{x}} \rangle_t) - f(\langle \hat{\mathbf{x}} \rangle_{t-1})| < \theta \tag{5.9}$$

becomes true, the mutation strength is increased via

$$\sigma = \sigma \cdot \tau \tag{5.10}$$

with $\tau > 1$. Otherwise, the mutation strength σ is decreased by multiplication with $1/\tau$.

An increasing mutation strength σ allows to leave local optima. Powell's method drives the search into local optima, and the outer ILS performs a search within the *space of local optima* controlling the perturbation strength σ. A decrease of step size σ lets the algorithm converge to the local optimum in a range defined by σ. This technique seems to be in contraposition to the 1/5th success rule by Rechenberg [15]. Running a simple $(1 + 1)$-ES with isotropic Gaussian mutations

---

Algorithm 3 Powell ES

 1: initialize μ solutions
 2: apply Powell's method
 3: **repeat**
 4:    **for** $j = 1$ **to** λ **do**
 5:       mutate solution $\mathbf{x}_j$
 6:       apply Powell's method
 7:    **end for**
 8:    select μ-best solutions
 9:    **if** fitness improvement $< \theta$ **then**
10:       $\sigma = \sigma \cdot \tau$
11:    **else**
12:       $\sigma = \sigma/\tau$
13:    **end if**
14: **until** termination condition

---

and constant mutation steps σ, the optimization process will become very slow after a few generations. Rechenberg's rule adapts the mutation strengths in the opposite kind of way. If the ratio g/G of successful generations g after G generations is larger than 1/5th, the step size should be increased. The increase is reasonable, because bigger steps towards the optimum are possible, while small steps would be a waste of time. If the success ratio is less than 1/5th, the step size should be decreased. This rule is applied every G generations. The goal of Rechenberg's approach is to stay in the *evolution window* guaranteeing nearly optimal progress. Optimal progress is problem-dependent and can be stated theoretically on artificial functions [16].

However, in our approach the strategy of Powell approximates local optima, not the evolution strategy. The step control of the Powell ES has another task: leaving local optima, when the search stagnates. Basins of attractions can be left because of the increasing step size. Hence, the probability of finding the global optimum is larger than 0. With this mechanism, also the global optimum may be left again. But if the vicinity of the optimum has been reached, it is probable that it will be successively reached again. The problem that the global optimum may be left, if not recognized, can be compensated by saving the best found solution in the course of the optimization process.

5.5 Experimental Analysis

In the following, we will experimentally analyze the Powell ES on a set of test problems, cf. Appendix A. Again, initial solutions are generated in the interval $[-100, 100]^N$, and the step sizes are set to $\sigma_{\text{init}} = 1.0$. Each experiment is repeated 30 times. For the Powell ES, we employ the settings $\lambda = 8$ and $\mu = 2$. Each solution is mutated and locally optimized with Powell's method. Again, Powell's method terminates, if the improvement from one to the next iteration is smaller than $\phi = 10^{-10}$, or if the optimum is found with accuracy $f_{\text{stop}} = 10^{-10}$. If the search on the ILS level stagnates, i.e., if the achieved improvement is smaller than θ, the mutation strength is increased with mutation parameter $\tau = 2.0$. We allow a maximal budget of $\text{ffe}_{max} = 2.0 \times 10^6$ fitness function evaluations.

Table 5.2 shows the results of the analysis of the Powell ES on the test problems with $N = 10$ and $N = 30$ dimensions. The results have shown that Powell's method is very fast on unimodal problems. Of course, the Powell ES shows the same capabilities and approximates the optimum in the first Powell-run on the *Sphere* problem and *Doublesum*. We have already observed that Powell gets stuck in local optima of multimodal problems (e.g. *Rastrigin*). The Powell ES perturbates a solution, when getting stuck, and applies Powell's method again with the perturbation mechanism of Eq. (5.10). The results show that the iterated application of Powell's method in each generation allows to approximate the global optimum, also on *Rastrigin*. The Powell ES is able to approximate the optimum in comparison to its counterpart without ILS.

It convergences significantly faster than the CMSA-ES (cf. Chap. 2). A statistically significant superiority of the Powell ES can also be observed on *Griewank*. On *Rosenbrock*, no superiority of any of the two algorithms can be reported. Although the worst runs of the Powell ES cause a fitness deterioration in mean, the best runs are still much faster than the best runs of the CMSA-ES. The CMSA-ES is more robust with smaller standard deviations, but does not offer the potential to find the optimal solution that fast. A similar behavior can be observed on the test problems with $N = 30$ dimensions, see the lower part of Table 5.2. The CMSA-ES takes about 17 times more evaluations. This also holds true for the other unimodal test problems, where the Powell ES is superior. On the multimodal test problems in higher dimensions, similar results as for $N = 10$ can be observed. The Powell ES

Table 5.2 Experimental analysis of the Powell ES on the test problems with $N = 10$ and $N = 30$ dimensions

	Best	Median	Worst	Mean	Dev	#
$N = 10$						
f_{Sp}	99	100	153	105.1	1.6e1	30
f_{Dou}	89	92	178	108.6	3.6e1	30
f_{Ros}	3,308	5,074.5	29,250	7,772.6	7.8e3	30
f_{Ras}	2,359	14,969.5	38,550	15,682.5	9.1e3	30
f_{Gri}	477	1,506	6,572	2,240	2.2e3	30
f_{Kur}	4,300	196,218	316,528	165,325	9.9e4	30
$N = 30$						
f_{Sp}	290	295.5	299	295.3	2.83	30
f_{Dou}	283	286.5	479	305.6	6.0e1	30
f_{Ros}	25,363	61,768	385,964	95,339	1.0e5	30
f_{Ras}	58,943	78,537.5	191,489	102,429	4.4e4	30
f_{Gri}	971	5,994.5	20,913	9,629.6	7.6e3	30
f_{Kur}	>2.0e6	>2.0e6	>2.0e6	>2.0e6	–	0

The figures show the number of fitness function evaluations until the difference between the fitness of the best solution and the fitness of the optimum is smaller than $f_{stop} = 10^{-10}$. This termination condition has been reached in every run except on f_{Kur} with $N = 30$.

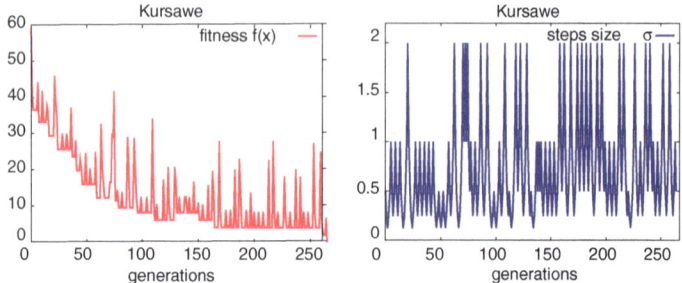

Fig. 5.1 Development of fitness and step sizes on the multimodal problem *Kursawe* employing $N = 10$. When the search gets stuck in local optima, the perturbation mechanism increases σ and lets the Powell ES escape from basins of attraction [1]

is statistically better on *Rastrigin*. The CMSA-ES's mean and median are better on *Rosenbrock* and *Griewank*. On *Kursawe*, the optimum has been found in every run for $N = 10$, but in no run for $N = 30$.

Figure 5.1 shows fitness curves and step sizes of typical runs on the multimodal problem *Kursawe* with $N = 10$. It can be observed that the search successively gets stuck. But the perturbation mechanism always allows to leave the local optima again. When the search gets stuck in a local optimum, the strategy increases σ until the local optimum is successfully left, and a better local optimum is found. The approach moves from one local optimum to another controlling σ, until the global optimum is found. The fitness development reveals that the search has to accept worse solutions to approximate the optimal solution. The figures confirm the basic idea of the algorithm. ILS controls the global search, while Powell's method drives the search into local optima. Frequently, the hybrid is only able to leave local optima by controlling the strength σ of the Gaussian perturbation mechanism. ILS conducts a search in the space of local optima.

5.6 Perturbation Mechanism and Population Sizes

For deeper insights into the perturbation mechanism and the interaction with population sizes, we conduct further experiments on the multimodal problem *Rastrigin* with $N = 30$, where the Powell ES has shown successful results. The strength of the perturbation mechanism plays an essential role for the ILS. In case of stagnation, the step size is increased as described in Eq. (5.10) with $\tau > 1$ to let the search escape from local optima. Frequently, a successive increase of the perturbation strength is necessary to prevent stagnation. In case of an improvement, the step size is decreased with $\tau < 1$. The idea of the step size reduction is to prevent the search process from jumping over promising regions of the solution space. In the following, we ana-

Table 5.3 Analysis of the Powell ES perturbation parameter τ and the population sizes on *Rastrigin* with $N = 30$ using the same initial settings, performance measure, and termination condition like in the previous experiments

(μ, λ)	Best	Median	Worst	Best	Median	Worst
	$\tau = 1.5$			$\tau = 2.0$		
(1, 4)	53,913	92,544	130,675	31,686	78,214	121,170
(2, 8)	56,074	100,835	143,642	65,540	112,643	242,983
(4, 16)	149,350	162,083	210,384	77,481	117,972	163,693
(8, 32)	156,517	295,457	370,320	193,259	209,725	244,325
	$\tau = 5.0$			$\tau = 10.0$		
(1, 4)	53,465	105,513	406,495	$>2 \times 10^9$	$>2 \times 10^9$	$>2 \times 10^9$
(2, 8)	48,274	104,461	285,651	32,773	680,363	1,473,097
(4, 16)	67,241	103,142	202,447	52,991	208,088	338,922
(8, 32)	109,820	189,676	221,069	123,838	309,169	802,285

lyze the perturbation mechanism and the population sizes on *Rastrigin*. We try to determine useful parameter settings for τ and for population parameters μ and λ.

Table 5.3 shows the corresponding results. The best result has been achieved with $\tau = 2.0$ and population sizes (1, 4). Also the best median has been achieved with this setting, while the second best has been achieved with $\tau = 1.5$ and population sizes (1, 4). With parameter setting $\tau = 10.0$, the Powell ES achieves a satisfying best solution, but the variance of the results is high. The worst solution is comparably bad. In general, the results for $\tau = 10.0$ are quite weak, for (1, 4) the algorithm does not converge within reasonable time. For low mutation strengths, the best results can be observed for small population sizes. In turn, for higher mutation strengths, i.e., $\tau = 5.0$, larger population sizes are necessary to compensate the explorative effect. Further experiments on other problems led to the decision that a (2, 8)-Powell ES is a good compromise between exploration and efficiency, while a (4, 16)-Powell ES is a rather conservative, but stable choice with reliable results.

5.7 Conclusions

Combining the world of local search with the world of global evolutionary optimization is a promising undertaking. It reflects the original idea of evolutionary computation. If we do not know anything about the problem, evolutionary algorithms are an appropriate choice. In multimodal fitness landscapes, we typically know nothing about the *landscape of local optima*. The Powell ES only assumes that attractive local optima lie closely together. Hence, the search might jump from one basin of attraction to a neighbored one. To move into local optima, Powell's method turns out to be fairly successful. Furthermore, the adaptation of the perturbation strength is a natural enhancement in real-valued solution spaces. A population-based implementation allows to run multiple Powell searches in parallel and to achieve a crucial speedup in distributed computing environments.

References

1. O. Kramer, Iterated local search with Powell's method: a memetic algorithm for continuous global optimization. Memetic Comput. **2**(1), 69–83 (2010)
2. C. Blum, M. Aguilera, A. Roli, M. Sampels, Hybrid Metaheuristics: an Introduction. In Hybrid Metaheuristics (Springer, Berlin, 2008), pp. 1–30
3. H.R. Lourenço, O. Martin, T. Stützle, *Iterated Local Search*, eds. by F. Glover, G. Kochenberger. In Handbook of Metaheuristics, (Kluwer Academic, Norwell, 2003), pp. 321–352
4. H.R. Lourenço, O. Martin, T. Stützle, A beginner's introduction to iterated local search, in *Proceedings of the Fourth Metaheuristics Conference*, vol. 2 (Porto, Portugal, 2001), pp. 1–6
5. T. Stützle, H.H. Hoos, Analyzing the run-time behaviour of iterated local search for the TSP, in *Proceedings III Metaheuristics International Conference* (Kluwer Academic Publishers, Norwell, 1999)
6. K. Katayama, H. Narihisa, Iterated local search approach using genetic transformation to the traveling salesman problem, in *Proceedings of the 1st Conference on Genetic and Evolutionary Computation (GECCO) ACM*, New York, USA, 1999
7. T. Stützle, Iterated local search for the quadratic assignment problem. Eur. J. Oper. Res. **174**(3), 1519–1539 (2006)
8. A.R. Duarte, C.C. Ribeir, A hybrid ILS heuristic to the referee assignment problem with an embedded MIP strategy, *Hybrid Metaheuristics* (2007), pp. 82–95, http://link.springer.com/chapter/10.1007%2F978-3-540-75514-2_7
9. N. Mladenovic, P. Hansen, Variable neighborhood search. Comput. Oper. Res. **24**, 1097–1100 (1997)
10. F. Glover, M. Laguna, *Tabu Search* (Springer, Berlin 1997)
11. A. Griewank, Generalized decent for global optimization. JOTA **34**, 11–39 (1981)
12. M. Powell, An efficient method for finding the minimum of a function of several variables without calculating derivatives. Comput. J. **7**(2), 155–162 (1964)
13. M.J.D. Powell, Restart procedures for the conjugate gradient method. Math. Program. **V12**(1), 241–254 (1977)
14. H.-P. Schwefel, *Evolution and Optimum Seeking. Sixth-Generation Computer Technology* (Wiley, New York, 1995)
15. I. Rechenberg, *Evolutionsstrategie: Optimierung technischer Systeme nach Prinzipien der biologischen Evolution* (Frommann-Holzboog, Stuttgart, 1973)
16. H.-G. Beyer, *The Theory of Evolution Strategies* (Springer, Berlin, 2001)

Chapter 6
Multiple Objectives

6.1 Introduction

In many design process scenarios, the optimization of more than one objective at once is a frequent problem. This setting is a particularly difficult task, when the optimization objectives are conflictive, i.e., minimization of one objective potentially results in maximization of another. Strategies that allow the optimization of two or more conflictive objectives at a time are based on evolving a set of non-dominated solutions, i.e., solutions that are better in at least one objective than the other solutions in the set. Evolutionary algorithms have proven to be very popular solution strategies for multi-objective optimization problems. One reason is that evolutionary algorithms are quite easy to implement and to use. The second reason is that a set-based optimization framework is very appealing, as we seek for a set of Pareto-optimal solutions. The is often not a single unique solution that is optimal, but a whole set of solutions.

Prominent examples for evolutionary multi-objective algorithms (EMOAs)[1] are NSGA-ii [1] and SMS-EMOA [2]. In this chapter, we present a heuristic for selection of non-dominated solutions that is based on rakes, which are reference lines in objective space [3]. After the formal introduction of a multi-objective optimization problem, we present related work and introduce the rake approach. The approach will be experimentally analyzed and compared on a set of reference problems.

6.2 Multi-Objective Optimization

Many planning and decision making problems involve multiple conflictive objectives that have to be optimized simultaneously, in operations research also known as multiple criteria decision making. The unconstrained multi-objective optimization

---

[1] EMOAs are also known as multi-objective evolutionary algorithms (MOEAs).

O. Kramer, *A Brief Introduction to Continuous Evolutionary Optimization*,
SpringerBriefs in Computational Intelligence,
DOI: 10.1007/978-3-319-03422-5_6, © The Author(s) 2014

problem in $\mathbb{R}^N$ to minimize m conflictive objectives $\mathbf{f}(\mathbf{x}) = (f_1(\mathbf{x}), \dots, f_m(\mathbf{x}))^T$ is defined as

$$\min_{\mathbf{x} \in \mathbb{R}^N} \mathbf{f}(\mathbf{x}) = \min_{\mathbf{x} \in \mathbb{R}^N} (f_1(\mathbf{x}), f_2(\mathbf{x}), \dots, f_m(\mathbf{x}))^T \tag{6.1}$$

with $f_i(\mathbf{x}) : \mathbb{R}^N \to \mathbb{R}$, $i = 1, \dots, m$. In most cases, the decision maker is interested in a set of Pareto optimal solutions. To define Pareto optimality, we need the following relations for solutions in multi-objective minimization scenarios. Solution $\mathbf{x}$ weakly dominates solution $\mathbf{x}'$, written as

$$\mathbf{x} \preceq \mathbf{x}', \tag{6.2}$$

if it holds

$$\forall i \in \{1, \dots, m\} : f(x_i) \leq f(x_i'). \tag{6.3}$$

Solution $\mathbf{x}$ dominates solution $\mathbf{x}'$, written as

$$\mathbf{x} \prec \mathbf{x}', \tag{6.4}$$

if it holds

$$\mathbf{x} \preceq \mathbf{x}' \text{ and } \exists i \in \{1, \dots, m\} : f(x_i) < f(x_i'). \tag{6.5}$$

The solutions are incomparable, written as $\mathbf{x} \| \mathbf{x}'$, if neither $\mathbf{x} \preceq \mathbf{x}'$ nor $\mathbf{x}' \preceq \mathbf{x}$ holds. With these definitions, Pareto optimality can be defined. We seek for a set of solutions that in objective space defines a Pareto front

$$\mathcal{PF} = \{\mathbf{f}(\mathbf{x}^*) \in \mathbb{R}^m | \nexists \mathbf{x} \in \mathbb{R}^N : \mathbf{x} \prec \mathbf{x}^*\}. \tag{6.6}$$

The corresponding set of solutions in the solution space is the Pareto set

$$\mathcal{PS} = \{\mathbf{x}^* \in \mathbb{R}^N | \mathbf{f}(\mathbf{x}^*) \in \mathcal{PF}\}. \tag{6.7}$$

Figure 6.1 shows the partitioning of objective space induced by a solution $\mathbf{x}$ for a minimization problem. The solution would be dominated by a solution in the blue part and would dominate solutions in the red part. The other two quadrants of objective space contain solutions that cannot be compared to $\mathbf{x}$, i.e., they are neither better nor worse.

Pareto optimal solutions are also known as non-inferior solutions. After a Pareto set has been generated, the decision maker can select the solutions that fit best to the preferences. Simple ways to handle multi-objective optimization problems are to concentrate on the most important objective, while treating all others as constraints or to aggregate all objectives to a composite function. The first alternative is based on first selecting the most important objective function, which is called preference function

$$f(\mathbf{x}) = f_i \text{ with } i \in \{1, \dots, m\}. \tag{6.8}$$

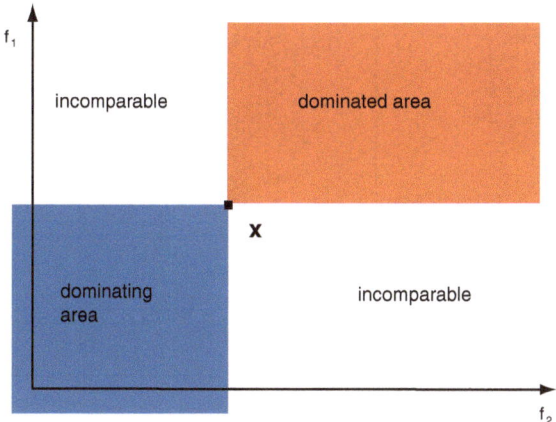

Fig. 6.1 Illustration of partitioning of a 2-dimensional objective space by a solution **x** for a minimization problem. Solutions in the *red area* are dominated by a solution **x**, while solutions in the *blue area* dominate solution **x**. Solutions in the *white part* of objective space are incomparable. The set of Pareto-optimal solutions contains incomparable solutions

For each remaining objective function, a constraint

$$g_j(\mathbf{x}) = f_j(\mathbf{x}) \text{ with } j \in \{1, \ldots, m\} \setminus i \tag{6.9}$$

is introduced. The composite method allows to handle multi-objective problems with single-objective methods without constraints. This idea will be combined with the rake idea in Sect. 6.6.

Research on multi-objective optimization has a long tradition beginning in the nineteenth century with the work of Edgeworth [4], Kuhn and Tucker [5], and Pareto [6]. EMOAs have shown outstanding success in the last decades. Algorithms like NSGA-ii by Deb et al. [1], SPEA by Zitzler and Thiele [7], and the SMS-EMOA by Emmerich et al. [8] are able to generate Pareto sets of solutions in non-linear and multimodal scenarios. Most EMOAs generate a population of non-dominated solutions to approximate the Pareto set. A comprehensive introduction to evolutionary multi-objective optimization is presented in the book by Coello'et al. [9].

Goldberg [10] was the first who introduced domination as selection objective. To maintain diversity, he introduced a niching-based approach. Also Horn et al. [11], as well as Fonseca and Fleming [12] use niching approaches. Many Pareto sampling techniques have been introduced: MOGA by Fonseca and Fleming [12], MOMGA, the multi-objective messy genetic algorithm by Veldhuizen and Lamont [13], MOMGA-ii by Zydallis et al. [14], and SPEA by Zitzler and Thiele [7], as well as its successor SPEA2 by Zitzler et al. [15]. One of the most famous approaches in this line of research is the non-dominated sorting genetic algorithm NSGA by Srinivas and Deb [16] and its successor NSGA-ii by Deb et al. [1]. The idea of non-dominated sorting is to rank solutions according to their non-domination level, see Sect. 6.3. An indicator in objective space is the basis for

Fig. 6.2 The dominated volume in objective space is called S-metric. The SMS-EMOA selects a set of solutions that maximizes the S-metric

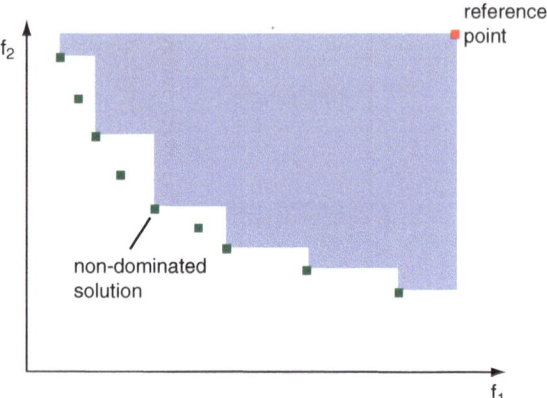

a second-level selection of NSGA-ii: best ranked solutions with maximum *crowding distance* values are added to the population of the next generation. The crowding distance selection is a diversity preserving mechanism in objective space. It is based on the average Manhattan distance to neighbored non-dominated solutions.

The S-metric is an indicator for the approximation of the Pareto front by computing the dominated hypervolume of a population. The metric has been introduced as basis of the SMS-EMOA. For an introduction, we refer to Emmerich et al. [8] and Beume et al. [2]. We will use the S-metric in the experimental analyses of this work as indicator for the ability to approximate the Pareto front. Figure 6.2 illustrates the S-metric that is defined as the part in objective space dominated by a set of non-dominated solutions w.r.t. a reference point.

6.3 Non-Dominated Sorting

In this section, non-dominated sorting is briefly introduced. All solutions that are not dominated are assigned to the first rank. In a next step, they are removed from the population, and the non-dominance check is conducted again. Both steps are repeated until all individuals have been assigned to a rank according to their non-domination level. The non-dominated subset of a population $\mathcal{P}$ is defined as

$$ND(\mathcal{P}) = \{\mathbf{x} \in \mathcal{P} | \nexists \mathbf{x}' \in \mathcal{P} \text{ with } \mathbf{x}' \prec \mathbf{x}\}. \tag{6.10}$$

The result of non-dominated sorting is a hierarchical partitioning of the population $\mathcal{P}$ into disjoint non-dominated sets $\mathcal{J}_1, \ldots, \mathcal{J}_v$ with

1. $\mathcal{J}_1 = ND(\mathcal{P})$,
2. $\forall i$ with $2 \leq i \leq v : \mathcal{J}_i = ND(\mathcal{P} \setminus \cup_{1 \leq j < i} \mathcal{J}_j)$,
3. $ND(\mathcal{J}_v) \setminus \mathcal{J}_v = \emptyset$,

with rank i. A similar result can be obtained, if we define that $\mathcal{J}_i$ contains all solutions that are dominated by i solutions. But solutions that are dominated by more than one solution, which do not dominate each other generate errors in comparison to the standard definition.

6.4 Rake Selection

If the decision maker wants to select solutions a posteriori, i.e., after the Pareto set has been generated, the question arises how the Pareto front should look like. Rake selection is an evolutionary multi-objective selection operator for selection of a subset of solutions from a set of non-dominated solutions presented by Kramer and Koch [3]. The idea is to define reference lines called rakes in objective space that define preference areas solutions are biased towards. Let μ be the number of rakes $l_1, \ldots, l_\mu$ in objective space. Let $ND(\{\mathbf{x}_1, \ldots, \mathbf{x}_\kappa\})$ be the set of non-dominated offspring solutions. For each rake l_i, rake selection chooses the closest solution from $ND(\{\mathbf{x}_1, \ldots, \mathbf{x}_\kappa\})$. Rakes can arbitrarily be placed in objective space. A reasonable placement is to distribute the rakes perpendicularly and equidistantly on a rake base, which is the line between the *corner points*. The corner points are solutions that are minimal w.r.t. each objective $\mathbf{c}_j = \mathbf{f}(\mathbf{x}_j^*)$ with $j = 1, \ldots, m$. The rakes cut the rake base in the intersection points $\mathbf{p}_i$ with $i = 1, \ldots, \mu - 2$ for $m = 2$. Algorithm 11 shows the pseudocode that generates the rake placement.

Algorithm 11 Rake Generation

1: minimize objectives $f_1(\mathbf{x}), \ldots, f_m(\mathbf{x})$
2: compute corner points $\mathbf{c}_1, \ldots, \mathbf{c}_m$
3: compute rake normal vector
4: compute intersection points $\mathbf{p}_i$

Figure 6.3 illustrates, how rakes are employed for the selection of non-dominated solutions. The blue dots and green squares are the non-dominated solutions. The red dots represent dominated solutions. The rakes are distributed on the rake base between the two corner points $\mathbf{c}_1$ and $\mathbf{c}_2$ intersecting the rake base in points $\mathbf{p}_1$–$\mathbf{p}_5$. For each rake, the closest non-dominated solution is selected, i.e., the green squares represent the candidate solutions that are finally chosen by rake selection.

Various experiments have shown that the exploration capabilities can be enhanced by shifting the rakes at the boundaries to the outside. Otherwise, the explorative character of the approach can be limited, as the rakes are only distributed between the corner points, and not the whole Pareto front may be covered. The outer rakes in the neighborhood of the corner points have an explorative character, while the inner points exploit the knowledge about the location of the already explored Pareto front.

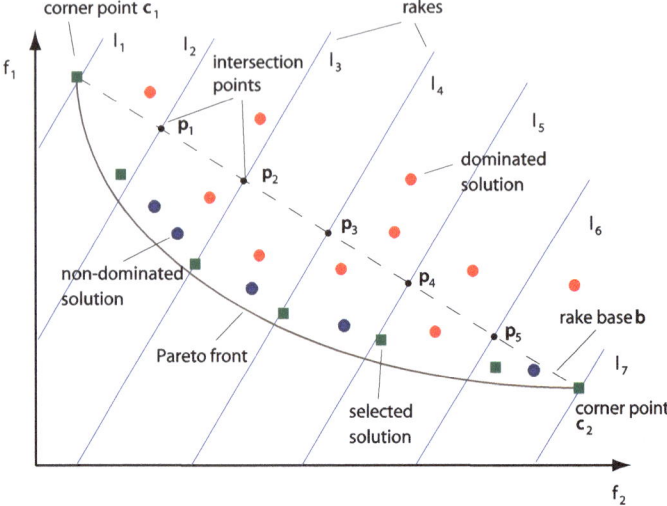

Fig. 6.3 Illustration of rake selection with parallel distribution, orthogonal to the line between the corner points for $m = 2$

Table 6.1 Experimental comparison between rake selection and SMS-EMOA in terms of S-metric on the multi-objective problems ZDT1, ZDT2, and ZDT6

Problem	EMOA	Best	Median	Worst	Dev
ZDT1	Rake	99.6558	99.65	99.64	0.003
	SMS	**99.6572**	99.65	99.65	0.0002
ZDT2	Rake	99.3233	99.32	99.31	0.003
	SMS	**99.3235**	99.32	90.00	3.901
ZDT6	Rake	**96.7411**	96.72	96.69	0.014
	SMS	95.6742	95.43	95.29	0.153

6.5 Experimental Study

In this section, we present an experimental study of the approach on the multi-objective ZDT test problems with $m = 2$ objectives. Minimization of the single objectives yield the corner points. We use $\mu = 50$ parental solutions, $k = 50$ rakes, and the mutation settings $\tau_0 = 5.0$ and $\tau_1 = 5.0$. Figure 6.4 shows the experimental results of typical runs of rake selection on the problems ZDT1 to ZDT4 with $N = 30$ dimensions ($N = 10$ for ZDT4) and random initialization after 1,000 iterations. We can observe that rake selection places the non-dominated solutions directly on the rakes and converges towards the Pareto front. The results are stable, i.e., the Pareto front is reached in almost every run.

We show the behavior of rake selection with a $(\mu + \lambda)$-ES and self-adaptive step sizes [17] on the ZDT problems ZDT1, ZDT2, and ZDT6 in Table 6.1. The

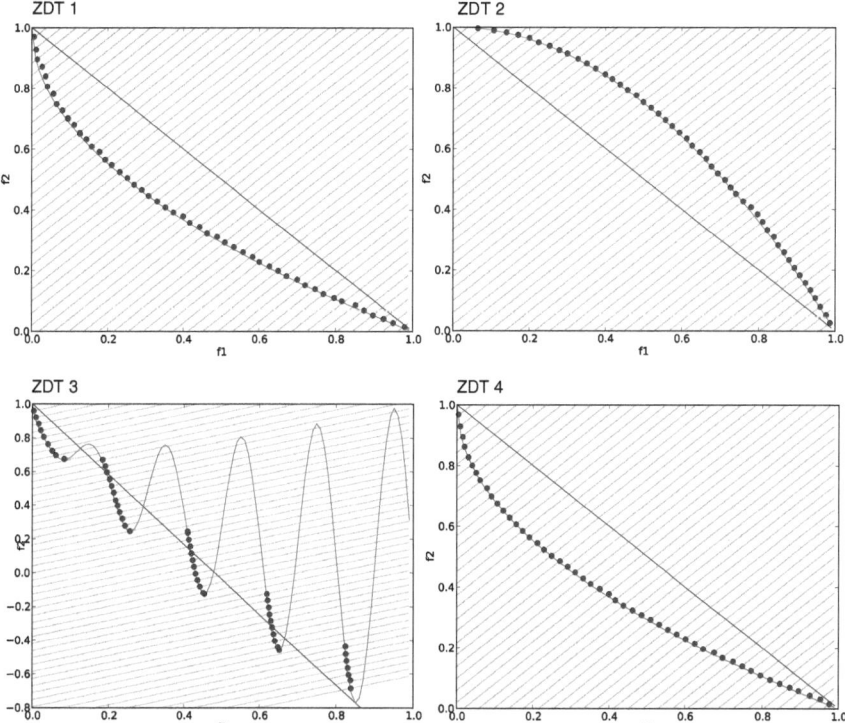

Fig. 6.4 Experimental results of typical runs of rake selection on the multi-objective problems ZDT1–ZDT4. The rake base connects the corner points. The perpendicular lines define the rakes in objective space. Due to different scalings of the axes, rake base and rakes do not appear to be orthogonal. After 1,000 iterations, i.e., 50,000 objective function evaluations, the solutions lie on the rakes and on the curves of the Pareto fronts

figures compare rake selection and SMS-EMOA w.r.t. the S-metric in 50 runs. We can observe that rake selection achieves high S-metric values. In case of ZDT6, rake selection even achieves higher values than the SMS-EMOA in average, although the fitness criterion of the SMS-EMOA is to explicitly maximize the S-metric.

6.6 Properties and Extensions

As the Pareto front is cone-convex [18], the condition of an approximately equidistant distribution will be fulfilled for most Pareto fronts. The solutions can easily be distributed equidistantly, if the Pareto front is linear. To improve the rake distribution in case of a strong curvature of the Pareto front, the intersection points $\mathbf{p}_i$ on the rake base can arbitrarily be adapted during the run, automatically or by hand.

Fig. 6.5 Result of ES on
ZDT1 with aggregated sum
(9,715 evaluations), see
Eq. 6.13

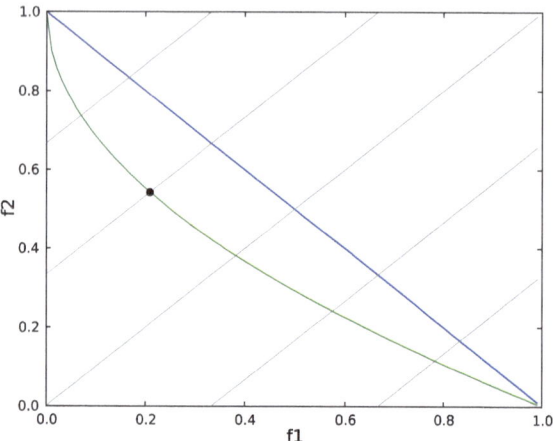

For the search with two objectives, rake selection requires μ rake lines depending on the density of solutions required for the Pareto front. To achieve a similar density of solutions in higher dimensional objective spaces with m objectives, μ^{m-1} rake lines are required. This exponential increase is a consequence of the *curse of dimensionality* in multi-objective optimization space. Neglecting that an equidistant distribution with constant distances into all directions of the objective space increases exponentially with the number of objectives leads to a loss of density in objective space.

In our experiments, the movement of solutions towards the Pareto front is achieved with non-dominated sorting. In case of the SMS-EMOA, the movement towards the Pareto front is achieved by maximizing the hypervolume in objective space w.r.t. to a dominated reference point $\mathbf{r}$ with

$$\forall \, \mathbf{x} \in \cup_{1 \le i \le t} \mathcal{P}_i : \mathbf{x} \prec \mathbf{r} \tag{6.11}$$

with population $\mathcal{P}_i$ at generation i. The concept of a reference point can also be used for rake selection by *maximizing* the distance to the reference point. This can be achieved by minimizing the weighted sum

$$\hat{f}(\mathbf{x}) = \alpha \cdot d(l^*, \mathbf{f}(\mathbf{x})) + (1 - \alpha)/\|\mathbf{f}(\mathbf{x}) - \mathbf{r}\|^2 \tag{6.12}$$

with weight $\alpha \in (0, 1)$ and $d(l^*, \mathbf{f}(\mathbf{x}))$ yielding the distance to the closest rake l^* and point $\mathbf{f}(\mathbf{x})$ in objective space. This mechanism was not necessary to achieve a movement towards the Pareto front of the ZDT-problems in the experimental analysis of Sect. 6.5, but has successfully been applied in further experiments.

To apply other single-objective optimizers, the aggregation of objectives may become necessary. Rake selection can be employed as summand for aggregating objectives in a weighted sum (composite) approach. Figure 6.5 illustrates, how rake selection can be combined with the weighted sum to approximate a single point on a line in objective space. Aggregating the distance to the closest rake $d(l^*, \cdot)$ with the weighted sum and weights $\mathbf{w} \in \mathbb{R}^m$, w' results in

$$\hat{f}(\mathbf{x}) = \mathbf{w}^T \mathbf{f}(\mathbf{x}) + w' \cdot d(l^*, \mathbf{f}(\mathbf{x})). \tag{6.13}$$

We employed an ES on ZDT1 with $N = 150$ dimensions. The result is a solution on the Pareto front with a distance of only $\approx 10^{-9}$ to the rake.

6.7 Conclusions

Rake selection is an evolutionary multi-objective algorithm based on non-dominated sorting. The distance to reference lines in objective space is used as indicator for the selection of Pareto optimal solutions. The reference lines can arbitrarily be placed in objective space. The selection is oriented to the distance to equidistantly distributed lines in objective space. Birth surplus allows the application of σ-self-adaptation. The experimental analysis has shown that the population follows the rakes and at the same times approximates the Pareto front. A weighted sum approach allows the employment of fast single-objective optimizers with rake selection. In [19], extensions of rake selection have been introduced that allow the approximation of equivalent Pareto subsets. The approach is based on the hybridization with clustering techniques and on mechanisms to trigger the clustering process.

References

1. K. Deb, S. Agrawal, A. Pratap, T. Meyarivan, A fast and elitist multi-objective genetic algorithm: NSGA-ii. IEEE Trans. Evol. Comput. **6**(2), 182–197 (2002)
2. N. Beume, B. Naujoks, M. Emmerich, SMS-EMOA: Multiobjective selection based on dominated hypervolume. Eur. J. Oper. Res. **181**(3), 1653–1669 (2007)
3. O. Kramer, P. Koch, Rake selection: a novel evolutionary multi-objective optimization algorithm, in *Proceedings of the German Annual Conference on Artificial Intelligence (KI)* (Springer, Berlin, 2009), pp. 177–184
4. F. Edgeworth, *Mathematical Psychics: An Essay on the Application of Mathematics to the Moral Sciences* (C. Kegan Paul & Co, London, 1881)
5. H.W. Kuhn, A.W. Tucker, *Nonlinear Programming* (1951), pp. 481–492
6. V. Pareto, Cours d'Economie Politique. Lausanne (1896)
7. E. Zitzler, L. Thiele, Multiobjective evolutionary algorithms: a comparative case study and the strength pareto approach. IEEE Trans. Evol. Comput. **3**(4), 257–271 (1999)
8. M. Emmerich, N. Beume, B. Naujoks, An EMO algorithm using the hypervolume measure as selection criterion, in *Evolutionary Multi-Criterion Optimization* (Springer, 2005), pp. 62–76
9. C.A. Coello Coello, G.B. Lamont, D.A. van Veldhuizen, *Evolutionary Algorithms for Solving Multi-Objective Problems*. Genetic and Evolutionary Computation Series, 2nd edn. (Springer Science+Business Media, New York, 2007)
10. D. Goldberg, *Genetic Algorithms in Search* (Optimization and Machine Learning, Addison-Wesley, Reading, MA, 1989)

11. J. Horn, N. Nafpliotis, D.E. Goldberg, A niched pareto genetic algorithm for multiobjective optimization, in *IEEE Conference on, Evolutionary Computation* (1994), pp. 82–87

12. C.M. Fonseca, P.J. Fleming, Genetic algorithms for multiobjective optimization: formulation, discussion, and generalization, in *International Conference on Genetic Algorithms* (1993), pp. 416–423

13. D.A. van Veldhuizen, G.B. Lamont, Multiobjective evolutionary algorithms: analyzing the state-of-the-art. Evol. Comput. **8**(2), 125–147 (2000)

14. J.B. Zydallis, D.A. van Veldhuizen, G.B. Lamont, A statistical comparison of multiobjective evolutionary algorithms including the MOMGA-ii, in *Evolutionary Multi-Criterion, Optimization* (2001), pp. 226–240

15. E. Zitzler, M. Laumanns, L. Thiele, SPEA2: Improving the Strength Pareto Evolutionary Algorithm for Multiobjective Optimization, in *Evolutionary Methods for Design, Optimisation, and Control with Application to Industrial Problems (EUROGEN)* (2002), pp. 95–100

16. N. Srinivas, K. Deb, Multiobjective optimization using nondominated sorting in genetic algorithms. Evolutionary Computation **2**, 221–248 (1994)

17. H.-G. Beyer, H.-P. Schwefel, Evolution strategies—A comprehensive introduction. Nat. Comput. **1**, 3–52 (2002)

18. S. Boyd, L. Vandenberghe, *Convex optimization* (Cambridge University Press, Cambridge, 2004)

19. O. Kramer, H. Danielsiek, A clustering-based niching framework for the approximation of equivalent pareto-subsets. Int. J. Comput. Intell. Appl. **10**(3), 295–311 (2011)

Part III
Learning

Chapter 7
Kernel Evolution

7.1 Introduction

In supervised learning scenarios, the objective is to learn a functional model f that best explains a set of observed patterns with their corresponding labels. Classification is the discrete and regression the continuous variant of this learning problem (cf. Chap. 1). The Nadaraya-Watson estimator [1, 2], also known as kernel (density) regression, is a famous regression method. Important parts of kernel regression are kernel density functions that estimate densities of patterns in data space. The densities are used as distribution estimates in the regression function formulation.

The model quality of kernel regression significantly depends on the choice of kernel parameters. Formulated as optimization problem evolution strategies are used for optimizing parameters of the Nadaraya-Watson estimator, which are used for searching in the space of kernel density parameters. They are integrated into a framework of leave-one-out cross-validation (LOO-CV) and arbitrary loss functions that can also be non-differentiable. An extension to local models with separate kernel parameters allows the adaptation to local data space characteristics.

7.2 Kernel Density Regression

Kernel density regression is a regression technique that has already been introduced in the sixties of the last century [1, 2]. With the success of kernel methods like SVMs, also kernel density regression got more attention, e.g., in unsupervised kernel regression [3]. Kernel regression weights the output values $y_i \in \mathbb{R}^d$ of patterns $x_i \in \mathbb{R}^q$, $i = 1, \ldots, N$ with relative kernel densities in input space. The idea has been introduced by Nadaraya and Watson [1, 2] and is known as *Nadaraya-Watson estimator*. It is based on kernel density estimation with kernel density functions, which share similarities with histograms. Figure 7.1 shows two histograms for a sample of 100 Gaussian points. Histograms count the number of samples that fall

O. Kramer, *A Brief Introduction to Continuous Evolutionary Optimization*,
SpringerBriefs in Computational Intelligence,
DOI: 10.1007/978-3-319-03422-5_7, © The Author(s) 2014

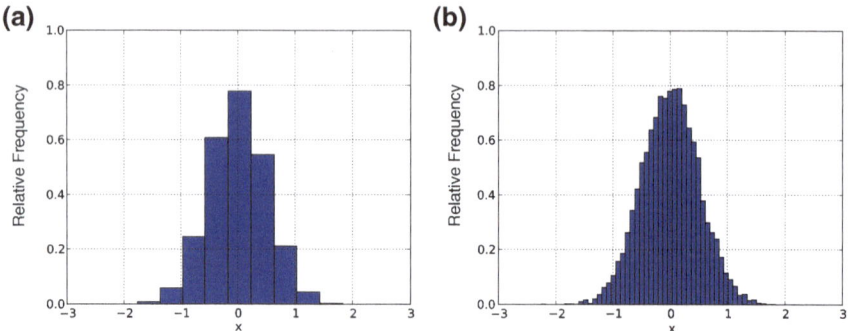

Fig. 7.1 Histograms count the number of points that fall into a bin of defined width. The larger the width the coarser-grained are the bins and consequently, the worse is the resolution of the histogram, **a** histogram, coarse bins, **b** histogram, fine bins

into intervals of defined widths. The length of a histogram bin is proportional to the number of points that fall into its bin. The influence of the histogram widths is illustrated. The larger the interval width, the coarser-grained are the bins and the worse is the resolution of the histogram. In kernel regression, each point is represented by a kernel function with a bump-like shape, e.g., a Gaussian function. The sum of the small bumps at each pattern delivers a continuous estimate of the distribution of points. The curvature of the bumps is defined by kernel functions.

Let $p(x, y) = (X = x, Y = y)$ be the joint distribution of patterns with distribution X and labels with distribution Y. The optimal mapping f^* is known as regression function and can be expressed in terms of the distributions $p(x, y)$ and $p(x)$ (cf. Bishop [4] and Hastie et al. [5])

$$f^*(x) = \int yp(y|x)dy = \int y\frac{p(x, y)}{p(x)}dy. \tag{7.1}$$

The idea of kernel regression is to express the joint distribution $p(x, y)$ with its kernel density estimate

$$\hat{p}(x, y) = \frac{1}{N}\sum_{i=1}^{N}K_{w_x}(x - x_i)K_{w_y}(y - y_i) \tag{7.2}$$

with a kernel density function K_{w_x} in data space and a kernel density function K_{w_y} in output space. This expression for the joint density $\hat{p}(x, y)$ and a similar expression for density $\hat{p}(x)$ inserted into Eq. 7.1 yield

$$\hat{f}(x) = \sum_{i=1}^{N}\frac{K_{w_x}(x - x_i)}{\sum_{j=1}^{N}K_{w_x}(x - x_j)}\int yK_{w_y}(y - y_i)dy \tag{7.3}$$

The solved integral is also known as Nadaraya-Watson estimator and weights the output values of the training samples with their relative kernel densities. The multi-variate Nadaraya-Watson estimator is defined as

$$\mathbf{f}(\mathbf{x}, \mathbf{W}) = \sum_{i=1}^{N} y_i \frac{K_\mathbf{W}(\mathbf{x} - \mathbf{x}_i)}{\sum_{j=1}^{N} K_\mathbf{W}(\mathbf{x} - \mathbf{x}_j)} \tag{7.4}$$

with matrix $\mathbf{W}$ containing the bandwidths. The bandwidth is an important parameter that controls the smoothness of the functional model. Small values lead to an overfit-ted prediction function, while high values tend to overgeneralize. The task to adapt the bandwidth has an important part to play. Let N be the number of patterns. For the prediction of one pattern label, N kernel densities have to be computed.

Kernel regression is based on a density estimate of patterns with a kernel function $K : \mathbb{R}^q \to \mathbb{R}$. In the experimental section, we will employ the Gaussian and the Epanechnikov kernel that are shortly introduced in the following. A typical kernel function is the multivariate Gaussian kernel

$$K_G(\mathbf{z}) = \frac{1}{(2\pi)^{q/2}\det(\mathbf{W})} e^{-\frac{1}{2}|\mathbf{W}^{-1}\mathbf{z}|^2} \tag{7.5}$$

with bandwidth matrix $\mathbf{W} = \mathrm{diag}(w_1, w_2, \ldots, w_q)$. Another frequent kernel is the Epanechnikov kernel

$$K_E(\mathbf{z}) = D_E \left(\frac{|\mathbf{z}|}{w} \right) \tag{7.6}$$

with

$$D_E(t) = \frac{3}{4}[1 - t^2]_+ = \begin{cases} \frac{3}{4} \cdot (1 - t^2) & |t| < 1 \\ 0 & |t| \geq 1 \end{cases} \tag{7.7}$$

Here, the bandwidth w defines the radius of the supported region, similar to the band-width of the Gaussian kernel. But unless the Gaussian kernel, it has a finite support and becomes 0 outside. For $h \to 0$, the Nadaraya-Watson estimator reconstructs the patterns, for $h \to \infty$, it averages the over all N patterns [6].

The result of the kernel density estimation significantly depends on the choice of proper kernel bandwidths. Small values lead to an overfitted prediction function, while high values result in an overgeneralization. LOO-CV is a technique to regu-larize the model. Various bandwidth selection methods are known in literature. A simple and good working choice is the Silverman's rule of thumb

$$w = \hat{\sigma} c \mu^{-1/5} \tag{7.8}$$

with μ solutions, the sample standard deviation $\hat{\sigma}$ and $c = 1.06$ for a Gaussian distribution, cf. Silverman [7]. The resulting choice of bandwidth w is known to be a good recommendation in many applications. A combination of LOO-CV and evolution strategies will be employed in the following.

In regression, typically different loss functions are used that weight the residuals. In the best case, the loss function is chosen according to the requirements of the data mining model and the application domain. With the design of a loss function, the emphasis of outliers can be controlled. Let $L : \mathbb{R}^d \times \mathbb{R}^d \to \mathbb{R}$ be the loss function. In the univariate case $d = 1$, a loss function is defined as $L = \sum_{i=1}^{N} L_k(y_i, f(\mathbf{x}_i))$. The L_1 loss is defined as $L_1 = |y - f(x)|$ and L_2 is defined as $L_2 = (y - f(x))^2$. Huber's loss [8] is a differential alternative to the L_1 loss and makes use of a trade-off point δ between the L_1 and the L_2 characteristic

$$L_h(r) = \begin{cases} \frac{1}{2 \cdot \delta} r^2 & |r| < \delta \\ |r| - \frac{1}{2}\delta & |r| \geq \delta \end{cases} \qquad (7.9)$$

with residual $r = y - f(x)$. Parameter δ allows a problem specific adjustment to certain problem characteristics. In the experimental part of this chapter, we employ Huber's loss with setting $\delta = 0.01$.

7.3 Kernel Shape Optimization

There are many examples in literature showing that evolutionary methods are successful in kernel-based machine learning. Stoean et al. [9, 10] directly solve the primal optimization problem of SVMs to find the optimal discriminant function for regression and classification tasks by means of evolution strategies. Mierswa and Morik [11] investigated simple data sets, where feature spaces induced by usual kernel functions fail. They employ a generic kernel learning scheme that is based on non-convex optimization. Furthermore, Mierswa [12] explicitly optimizes the tradeoff between training error and model complexity of SVMs by means of multi-objective evolutionary algorithms, i.e. NSGA-ii [13]. An example for the application of evolutionary methods to combinatorial problems in kernel-based machine learning stems from Gieseke et al. [14]. They solve the combinatorial problem of assigning elements to proper clusters with a $(1 + 1)$-EA. The approach aims at finding an optimal partitioning of data into two classes, which can be formulated as mixed integer problem. With the help of a kernel matrix approximation shortcut, computational costs can be reduced during approximation, and the evaluation of a huge set of solutions is possible within reasonable time. In our preliminary work, we used evolution strategies to adapt kernel parameters for the Nadaraya-Watson estimator and fit local models to energy consumption data [15].

Basis of the kernel shape optimization approach is a parameterized kernel density function that will be introduced in the following. To avoid overfitting, an LOO-CV scheme is employed. In this chapter, we have introduced standard kernel functions that are typically used for the Nadaraya-Watson estimator. To increase the flexibility, we extend the kernel functions to hybrid parameterized kernels that allow a greater flexibility to adapt to local data space characteristics. Bishop [4] states valid

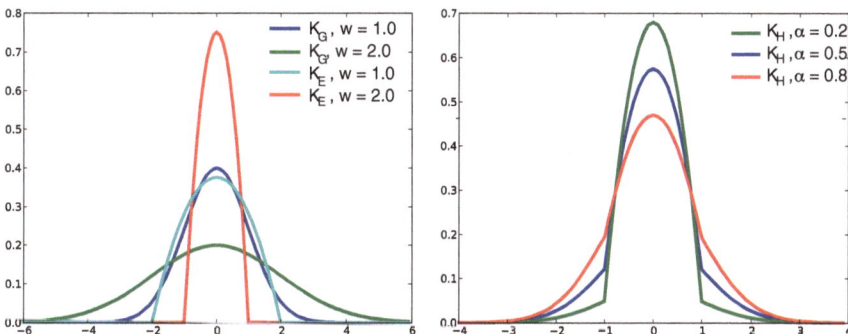

Fig. 7.2 Illustration of kernel density functions. *Left* Gaussian kernel K_G and Epanechnikov kernel K_E with two bandwidths $w = 1.0, 2.0$, *right* hybrid kernel K_H with $\alpha = 0.2, 0.5$, and 0.8

combinations of kernel functions. The following kernel density function, which is the weighted sum of the Gaussian and the Epanechnikov kernel

$$K_H(\mathbf{z}) = \alpha K_G(\mathbf{z}) + (1 - \alpha) K_E(\mathbf{z}) \tag{7.10}$$

with $\alpha \in [0, 1]$, is used. The novel kernel function can *morph* between both kernel density functions with parameter α. Figure 7.2 shows the shape of the novel hybrid kernel in comparison to the Gaussian and the Epanechnikov kernel in $q = 1$ for three settings of parameter α. For example, this can be useful, if only a limited support is necessary, and a Gaussian-similar shape is required.

If the model is trained w.r.t. the bandwidth diagonal matrix $\mathbf{W}$ on the training set, it may overadapt to the training examples and loose the ability to generalize. This effect called overfitting is likely to happen, if the training set size is small, or if the number of free parameters of a model is comparatively large. To avoid overfitting, Clark[16] selects the bandwidth matrix $\mathbf{W}$ as result of LOO-CV. The idea of LOO-CV is to apply the Nadaraya-Watson estimator, leaving out the pattern $(\mathbf{x}_i, \mathbf{y}_i)$ for each summand. The resulting error function that has to be minimized is

$$E = \frac{1}{N} \sum_{i=1}^{N} \|\mathbf{y}_i - \mathbf{f}_{-i}(\mathbf{x}_i, \mathbf{W})\|^2. \tag{7.11}$$

Here, $\mathbf{f}_{-i}$ denotes the Nadaraya-Watson estimator leaving out the i-th pattern. All points, with exception of data sample $(\mathbf{x}_i, \mathbf{y}_i)$ itself, contribute to the estimation of $\mathbf{f}(\mathbf{x}_i)$. For kernel shape optimization, i.e., choice of bandwidths and optimization of other kernel parameters, the CMA-ES is applied as alternative to grid search and Silverman's rule of thumb. As population sizes, the settings $\mu = 8$ and $\lambda = 16$ are chosen. The CMA-ES terminates after 100 fitness function evaluations, which is the same budget like grid search to allow a fair comparison.

Table 7.1 Analysis of search for optimal kernel parameters with Silverman's rule of thumb, grid search, and CMA-ES

Data	d	Silver		Grid		CMA-ES		
		K_G	K_E	K_G	K_E	K_G	K_E	K_H
f_{Sp}	8	0.080	>10.0	0.079	0.079	*0.072*	0.073	**0.070**
f_{Sp}	16	0.114	>10.0	0.094	0.094	**0.091**	0.093	**0.091**
f_{Sp}	32	0.134	>10.0	0.114	0.114	*0.113*	>10.0	**0.112**
f_{Dou}	8	0.160	>10.0	0.165	0.155	*0.141*	0.143	**0.135**
f_{Dou}	16	0.327	>10.0	0.280	0.280	**0.264**	0.276	*0.271*
f_{Dou}	32	0.614	>10.0	0.518	0.518	**0.461**	>10.0	*0.484*
f_{Ros}	8	1.156	>10.0	1.144	1.086	1.062	*1.033*	**1.032**
f_{Ros}	16	1.528	>10.0	1.353	1.371	*1.317*	1.330	**1.307**
f_{Ros}	32	2.185	>10.0	1.787	1.788	**1.740**	1.873	*1.752*
f_{Ras}	8	0.453	>10.0	0.356	0.356	**0.347**	*0.348*	0.352
f_{Ras}	16	0.585	>10.0	0.480	0.480	*0.462*	0.467	**0.455**
f_{Ras}	32	0.655	>10.0	0.549	0.549	*0.541*	0.597	**0.539**
f_{Gri}	8	0.032	>10.0	0.031	0.031	*0.022*	0.0221	**0.021**
f_{Gri}	16	0.034	>10.0	0.030	0.030	*0.023*	0.024	**0.021**
f_{Gri}	32	0.035	>10.0	0.028	0.028	*0.023*	0.030	**0.024**

7.4 Experimental Analysis

In the following, we analyze evolutionary kernel regression experimentally on a set of test functions, cf. Appendix A. The data sets have been created by sampling 100 patterns randomly with uniform distribution on the test problems with $q = 8, 16, 32$ dimensions in the unit cube $x_i \in [-1, 1]^q$, $i = 1, \ldots, 100$. We compare the introduced parameterized kernel density function in optimization scenarios employing Silverman's bandwidth rule, grid search, and the CMA-ES.

Table 7.1 shows the experimental results. For the evolutionary methods, the median of 50 runs is presented. The best results of each row are marked in bold, the second best in italic numbers. We can observe that the CMA-ES with hybrid kernel achieves the best LOO-CV error in 11 of the 15 cases. Consequently, the application of a hybrid kernel employing the CMA-ES can be recommended. The CMA-ES achieves the best result with Gaussian kernel in four cases. With one exception, grid search is always better than Silverman's rule of thumb. But Silverman's rule does not require any search (only one evaluation) and achieves significantly lower training errors than random parameterization. Silverman with Epanechnikov kernel fails, i.e., LOO-CV errors have been achieved that are much worse than the other results (>10.0). This is probably due to the fact that too low bandwidths are generated and can also be observed twice in the case of the CMA-ES optimization approach. Further, the CMA-ES with Epanechnikov kernel tends to be better than grid search with Epanechnikov kernel.

7.5 Local Models

In the previous kernel regression model, one global bandwidth matrix $\mathbf{W}$ has been used for the whole data space. This section enhances evolutionary kernel regression by the concept of local models that allow the adaptation of machine learning methods to local data space characteristics and as a consequence, results in high prediction precision in local data space areas [15, 17]. For each model, kernel density function parameters can be optimized w.r.t. local loss functions.

To handle varying data space characteristics at different data space regions, we present the concept of locality of the kernel regression methods based on codebook vectors. Each Nadaraya-Watson model is defined by a codebook vector $\mathbf{c}_k \in \mathbb{R}^q$, $k = 1, \ldots, K$ from the set of codebook vectors $\mathcal{C} = \{\mathbf{c}_1, \ldots, \mathbf{c}_K\}$. A pattern $(\mathbf{x}_i, \mathbf{y}_i)$, $i = 1, \ldots, N$ is assigned to the Nadaraya-Watson model $\mathbf{f}_{k^*}(\mathbf{x}_i, \mathbf{W}_{k^*})$ with minimal distance to its codebook vector k^*. The purpose of local Nadaraya-Watson models is to allow independent bandwidths and kernel parameterizations for increasing the adaptation flexibility in data space. For initialization, codebook vectors are distributed in data space according to the local kernel density. They are placed in regions with high kernel densities, but with a least distance to neighbored previously found regions with high kernel densities. For this sake, the patterns with the highest relative kernel density

$$kd(\mathbf{x}_j) = \sum_{i=1, i \neq j}^{N} K_{\mathbf{W}}(\mathbf{x}_i - \mathbf{x}_j) > \epsilon \qquad (7.12)$$

are identified iteratively with a minimum kernel density $\epsilon \in \mathbb{R}^+$ and a minimum distance $\rho \in \mathbb{R}^+$ to the previously computed codebook vectors $\mathbf{c}_k \in \mathcal{C}$ with $d(\mathbf{x}_j, \mathbf{c}_k) > \rho$. These points are added to the set $\mathcal{C}$ of codebook vectors. The initialization procedure is similar to kernel density clustering [18, 19], as it is based on the assignment of each point to the closest local optimum of the kernel density estimate [18, 19].

After initialization, the training phase starts. During training, the parameters of the local models are adapted. The optimization goal is to minimize the mean squared error of all local models w.r.t. the codebook vector set $\mathcal{C}$ and the local bandwidth matrices $\mathbf{W}_k$

$$E_{\text{local}} = \frac{1}{N} \sum_{i}^{N} \|\mathbf{y}_i - \mathbf{f}_{k^*}(\mathbf{x}_i, \mathbf{W}_{k^*})\|^2, \qquad (7.13)$$

with

$$k^* = \arg \min_{k=1, \ldots, K} \|\mathbf{x}_i - \mathbf{c}_k\|^2. \qquad (7.14)$$

As a side effect, the assignment of patterns to local models results in a significant speedup. If we assume that the patterns are uniformly distributed to all K models, which is only an idealized assumption, the computation time for the prediction of one pattern can be reduced to $1/K$-th. This also holds for the computation of the LOO-CV.

Table 7.2 Comparison between local evolutionary kernel regression (local EKR) and kernel regression (KR) employing grid search on a test data set with varying data densities

Regression	Train	Test	$Local_1$	$Local_2$
KR	0.4449	2.1397	2.1260	4.1634
Local EKR	0.4159	**2.0615**	1.9873	4.0694

Varying data densities and noise might afford separate parameterizations [15, 17]. First, we concentrate on local data densities. For this sake, we consider a trigonometric function $f(\mathbf{x}) = x_1 \cdot \cos(x_1) \sin(x_2)$ that consists of $K = 2$ uniformly at random distributed data clouds with $q = 5$ and $d = 1$. Each cloud consists of $N_k = 100$ patterns. While the first cloud consist of patterns that are uniformly at random generated in the interval $[0.0, 1.0]$, the second cloud is generated in the interval $[100.0, 110.0]$. The function value is not disturbed with noise. The resulting different pattern densities afford separate parameterizations. Table 7.2 confirms this expectation. It shows the comparison of training and test errors from local evolutionary kernel regression and kernel regression with grid search for w in the interval $[0, 100]$ in steps of 0.01.

For the codebook vector initialization with kernel density clustering, we use the least kernel density $\epsilon = 10^{-20}$ and the radius $\rho = 200$ to identify the set of model centers $\mathcal{C}$. Each local bandwidth matrix is initialized with diagonal entries $w_i = 20.0, i = 1, \ldots, q$. We use a population size of $\lambda = 4 \cdot N$ depending on the problem dimension N. It can be observed that local EKR finds two model centers that lie within the clouds. Local EKR achieves a lower training LOO-CV error using Huber's loss and also lower test errors on the whole test sets and for each local model.

An analysis of local EKR with varying noise magnitudes is the second experimental study we perform [15, 17]. The trigonometric function of the last section is used with the same data density, but different noise magnitudes disturbing the function value. Again, we generate two clouds of patterns, each consisting of $N_k = 100$ patterns. Noise can be modeled by adding Gaussian distributed random numbers to the original function value

$$\hat{y}_i = y_i + \alpha_j \cdot \mathcal{N}(0, 1) \tag{7.15}$$

with noise magnitude α_j in the j-th cloud. We set $\alpha_1 = 0.01$ for the first cloud and $\alpha_2 = 1.0$ for the second one. For Huber's loss function, we use the setting $\delta = 0.01$. Table 7.3 shows the corresponding experimental results. The local EKR approach is again able to achieve a lower LOO-CV error in the training phase than KR. Also the corresponding local test errors $local_1$ and $local_2$, as well as the overall test error are smaller than the results achieved by KR.

Table 7.3 Comparison between local EKR and KR with grid search on a test data set with varying noise magnitudes

Regression	Train	Test	$Local_1$	$Local_2$
KR	0.9844	18.6759	2.2763	38.0403
Local EKR	0.9658	**17.7813**	2.0840	36.2236

7.6 Conclusions

The choice of kernel density functions and kernel bandwidths is a multimodal optimization problem that we solved with the CMA-ES. The approach evolves kernel shapes with a flexible parameterized kernel density function and LOO-CV. We have compared the LOO-CV error of the CMA-ES to the LOO-CV error of other common approaches like grid search with the same budget of function calls and Silverman's rule of thumb. In the majority of the test cases, the CMA-ES with hybrid kernel turned out to be the best optimization algorithm. The CMA-ES is a good choice for parameter tuning of the Nadaraya-Watson estimator. The experiments on simple functions show that the application of hybrid kernels is recommendable. Furthermore, the approach allows the application of arbitrary, also non-differentiable kernel density functions. We could demonstrate that the application of local models with independent parameterizations is a recommendable approach to handle data space conditions like varying noise and pattern densities.

References

1. E. Nadaraya, On estimating regression. Theor. Probab. Appl. **10**, 186–190 (1964)
2. G. Watson, Smooth regression analysis. Sankhya Ser. A **26**, 359–372 (1964)
3. S. Klanke, H. Ritter, Variants of unsupervised kernel regression: general cost functions. Neurocomputing **70**(7–9), 1289–1303 (2007)
4. C. M. Bishop, *Pattern Recognition and Machine Learning (Information Science and Statistics)* (Springer, Berlin, 2007)
5. T. Hastie, R. Tibshirani, J. Friedman, *The Elements of Statistical Learning* (Springer, Berlin, 2009)
6. W. Härdle, L. Simar, *Appplied Multivariate Statistical Analysis* (Springer, Berlin, 2007)
7. B. W. Silverman, *Density Estimation for Statistics and Data Analysis, volume 26 of Monographs on Statistics and Applied Probability* (Chapman and Hall, London, 1986)
8. P.J. Huber, *Robust Statistics* (Wiley, New York, 1981)
9. R. Stoean, M. Preuss, C. Stoean, D. Dumitrescu, Concerning the potential of evolutionary support vector machines, in *IEEE Congress on Evolutionary Computation (CEC)* (2007), pp. 1436–1443
10. R. Stoean, M.P.C. Stoean, E. E-Darzi, D. Dumitrescu, Support vector machine learning with an evolutionary engine. J. Oper. Res. Soci. **60**(8), 1116–1122 (2009)
11. I. Mierswa, K. Morik, About the non-convex optimization problem induced by non-positive semidefinite kernel learning. Adv. Data Anal. Classif. **2**(3), 241–258 (2008)
12. I. Mierswa, Controlling overfitting with multi-objective support vector machines, in *Proceedings of the 9th Conference on Genetic and Evolutionary Computation (GECCO)* (ACM Press, New York, 2007), pp. 1830–1837
13. K. Deb, A. Pratap, S. Agarwal, T. Meyarivan, A fast and elitist multiobjective genetic algorithm: NSGA-II. IEEE Trans. Evol. Comput. **6**(2), 182–197 (2002)
14. F. Gieseke, T. Pahikkala, O. Kramer, Fast evolutionary maximum margin clustering, in *Proceedings of the International Conference on Machine Learning (ICML)* (ACM Press, New York, 2009), pp. 361–368
15. O. Kramer, B. Satzger, J. Lässig, Power prediction in smart grids with evolutionary local kernel regression, in *Hybrid Artificial Intelligence Systems (HAIS)* LNCS (Springer, Berlin, 2010), pp. 262–269

16. R. Clark, A calibration curve for radiocarbon dates. Antiquity **46**(196), 251–266 (1975)
17. O. Kramer, F. Gieseke, Evolutionary kernel density regression. Expert Syst. Appl. **39**(10), 9246–9254 (2012)
18. A. Hinneburg, D.A. Keim, A general approach to clustering in large databases with noise. Knowl. Inf. Syst. **5**(4), 387–415 (2003)
19. P. Schnell, A method to find point-groups. Biometrika **6**, 47–48 (1964)

Chapter 8
Particle Swarm Embeddings

8.1 Introduction

In big data scenarios, large numbers of high-dimensional patterns have to be processed. Efficient dimensionality reduction (DR) methods are required for algorithms that can only handle low-dimensional data like weak classifiers. With increasing data sets, DR methods becomes an important problem class in machine learning. Surprisingly, not many swarm-based algorithms for DR have been introduced in the past. DR methods compute a mapping from high-dimensional data space to a latent space of lower dimensionality. Latent point in this space should preserve the topological characteristics of their high-dimensional counterparts like neighborhood and distance relations. This chapter presents a novel iterative swarm-inspired approach for DR tasks. The particle swarm embedding algorithm (PSEA) combines the iterative construction of solutions with PSO equations. PSO is inspired by the movement of swarms in nature like fish schools or flocks of birds, and simulates the movement of candidate solutions using flocking-like equations with locations and velocities [1, 2]. The experimental part shows that PSEA is a powerful DR method.

8.2 Related Work

The idea of DR methods is to learn low-dimensional representations of high-dimensional patterns losing as little information as possible. Many DR methods seek for a mapping $\mathbf{F} : \mathbb{R}^d \to \mathbb{R}^q$ from a high-dimensional data space $\mathbb{R}^d$ to a latent space of lower dimensionality $\mathbb{R}^q$ with $q < d$. Non-parametric dimensionality reduction methods compute a set of low-dimensional representations $\mathbf{X} = (\mathbf{x}_1, \ldots, \mathbf{x}_N) \in \mathbb{R}^{q \times N}$ for N high-dimensional observed patterns $\mathbf{Y} = (\mathbf{y}_1, \ldots, \mathbf{y}_N) \in \mathbb{R}^{d \times N}$.

The decision, which information can be lost, and which has to be preserved in the mapping $\mathbf{F}$ depends on the purpose of the DR process and the error function defined for the employed method. Many DR methods use an implicit definition of

O. Kramer, *A Brief Introduction to Continuous Evolutionary Optimization*,
SpringerBriefs in Computational Intelligence,
DOI: 10.1007/978-3-319-03422-5_8, © The Author(s) 2014

the optimization problem they solve. However, the problem to learn the functional model $\mathbf{F}$ can be a hard optimization problem, because the latent variables $\mathbf{X}$ are unknown. Learning a reconstruction mapping $\mathbf{f} \colon \mathbb{R}^q \to \mathbb{R}^d$ back from latent to data space can also be desirable. Some methods learn this mapping automatically. Famous DR methods for non-linear dimensionality reduction are linear embedding (LLE) [3] and isometric mapping (ISOMAP) [4].

The framework for unsupervised regression has been introduced by Meinicke [5]. It is based on optimizing latent variables to reconstruct high-dimensional data. Unsupervised regression has first been applied to kernel density regression [6] and later to Gaussian processes [7] and neural networks [8]. Recently, we fitted nearest neighbor regression to the unsupervised regression framework [9], and introduced extensions w.r.t. robust loss functions [9]. Unsupervised nearest neighbors (UNN) is a fast approach that allows to iteratively construct low-dimensional embeddings in $\mathcal{O}(N^2)$, and has been introduced for latent sorting [9]. The approach we introduce in this work extends UNN with a PSO-like mechanism to handle arbitrary latent dimensionalities, i.e., $1 \leq q < d$. An introduction to UNN will be given in Sect. 8.3.

In nature, systems can be observed, in which comparatively simple units organize in groups. This form of collective and coordinated organization is known as swarm intelligence. The disadvantage of simple behaviors is compensated by their large number and massive parallelism. Swarms consist of a large number of simple entities that cooperate to act goal-oriented. Natural and artificial system have shown to implement successful solution strategies. To the best of our knowledge, no swarm-based methods have yet been proposed for embedding of patterns in low-dimensional latent spaces. But related work in other fields of unsupervised learning with swarm methods has been published, e.g., methods for PSO and ant colony optimization-based clustering. Kao and Cheng [10] have introduced an ACO algorithm for clustering that employs pheromones and distances between elements as heuristic clustering information. The combination of population-based search and stochastic elements allows to overcome local optima, and find optimal clustering results. Further methods for swarm-based clustering can be found in the book by Abraham et al. [11]. O'Neill and Brabazon [12] have introduced a hybrid approach of PSO, and self-organizing maps (SOMs) by Kohonen [13] that control the weights of a SOM employing a PSO-similar update rule. Also ant colony optimization has been employed to improve the topographic SOM mapping [14].

8.3 Iterative Particle Swarm Embeddings

PSEA combine K-nearest neighbor regression with the concept of unsupervised regression. The problem is to predict labels $\mathbf{y} \in \mathbb{R}^d$ to given patterns $\mathbf{x} \in \mathbb{R}^q$ based on sets of N pattern-label examples $(\mathbf{x}_1, \mathbf{y}_1), \ldots, (\mathbf{x}_N, \mathbf{y}_N)$. The goal is to learn a functional model $\mathbf{f} \colon \mathbb{R}^q \to \mathbb{R}^d$ known as regression function. We assume that a data set consisting of observed pairs $(\mathbf{x}_i, \mathbf{y}_i) \in \mathbf{X} \times \mathbf{Y}$ is given. For a novel pattern $\mathbf{x}'$, KNN regression computes the mean of the function values of its K-nearest patterns, see Eq. 4.7. The idea of KNN is based on the assumption of locality in data space: In local neighborhoods of $\mathbf{x}$ patterns are expected to have similar label information $\mathbf{f}(\mathbf{x})$ like observed patterns $\mathbf{y}$. Consequently, for an unknown $\mathbf{x}'$ the label must be similar to the labels of the closest patterns, which is modeled by the average of the output value of the K nearest samples. KNN has been proven well in various applications, e.g., in the detection of quasars based on spectroscopic data [15]. We define the output of function $\mathbf{f}_{KNN}$ given the pattern matrix $\mathbf{X}$ as a matrix $\mathbf{f}_{KNN}(\mathbf{X}) = [\mathbf{f}_{KNN}(\mathbf{x}_1), \ldots, \mathbf{f}_{KNN}(\mathbf{x}_N)]$, collecting all KNN mappings from patterns in $\mathbf{X}$ to $\mathbb{R}^d$.

The concept of unsupervised regression [5] is based on mapping from latent space to data space. The latent variables are the free parameters that have to be optimized to reconstruct the observed patterns in data space. Hence, the objective is to minimize the data space reconstruction error (DSRE):

$$\text{minimize } E(\mathbf{X}) = \frac{1}{N} \|\mathbf{Y} - \mathbf{f}_{KNN}(\mathbf{X})\|_F^2, \tag{8.1}$$

with Frobenius norm $\| \cdot \|_F^2$. We define $e(\mathbf{x}, \mathbf{y}, \mathbf{X})$ as the contribution of latent position $\mathbf{x}'$ to the DSRE

$$e(\mathbf{x}, \mathbf{y}, \mathbf{X}) = \|\mathbf{y} - \mathbf{f}_{KNN}(\mathbf{x}')\|^2. \tag{8.2}$$

The question comes up how to optimally place the latent positions. Figure 8.1a illustrates the unsupervised regression variant we proposed for sorting high-dimensional data [9]. It shows the $\hat{N} + 1$ possible embeddings of a data sample into an existing order of points in latent space (yellow/bright circles). The position of element $\mathbf{x}_3$ results in a lower DSRE with $K = 2$ than the position of $\mathbf{x}_5$, as the mean of the two nearest neighbors of $\mathbf{x}_3$ is closer to $\mathbf{y}$ than the mean of the two nearest neighbors of $\mathbf{x}_5$. Figure 8.1b shows an example of a UNN embedding of the 3D-S (upper part shows colorization of the *unsorted* S, lower part after UNN embedding), similar colors correspond to neighbored positions in latent space, i.e., a meaningful neighborhood preserving embedding has been computed. In the following, we extend the approach to arbitrary latent dimensionalities, in which the latent variables can be placed in latent space without geometric constraints.

There are two reasons to employ a direct search method to solve the UNN optimization problem. First, the problem is highly multimodal, second, $E(\mathbf{X})$ is not steady, and not differentiable due to the employment of KNN. To illustrate the search for optimal latent positions, we visualize the DSRE space in Fig. 8.2. It shows the DSRE w.r.t. the first pattern $\mathbf{y}_1$ for two neighborhood sizes, i.e., $K = 5$ (left), and $K = 30$ (right) after a run of UNN with $N = 300$. Bright areas represent parts of latent space with low errors, while dark colors represent a large DSRE. The comparison of both figures shows that in case of increasing neighborhood sizes the problem has larger, but less areas with similar fitness. The number of local optima decreases, and the optimization problem becomes easier. In the experimental section we will observe that the variance of the outcome of multiple experiments is smaller for large neighborhood sizes.

(a)   **(b)**

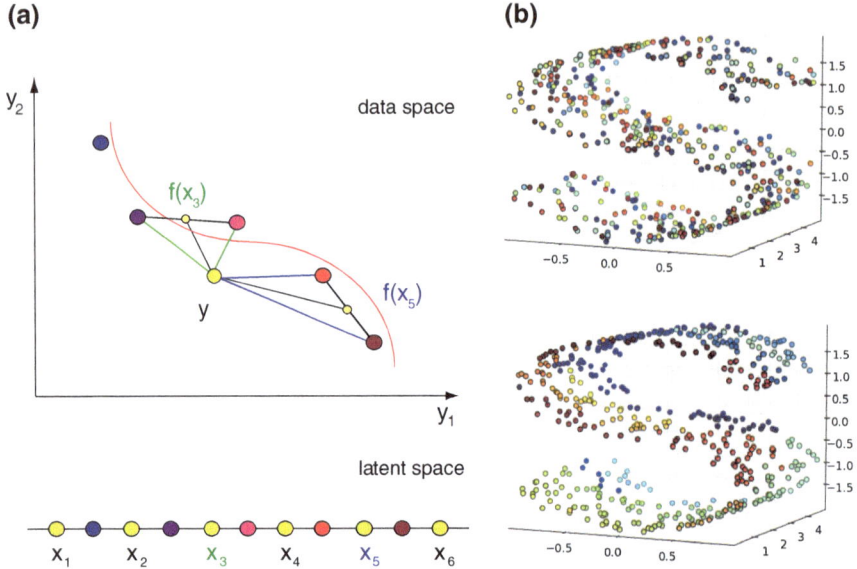

Fig. 8.1 *Left* Illustration of UNN embedding of a low-dimensional point to a fixed latent space topology testing all $\hat{N} + 1$ positions. *Right* Example of UNN result of a 3D-*S* before (*upper right*) and after embedding (*lower right*) with UNN and $K = 10$ [9]

(a) DSRE space, $K = 5$ **(b)** DSRE space, $K = 30$

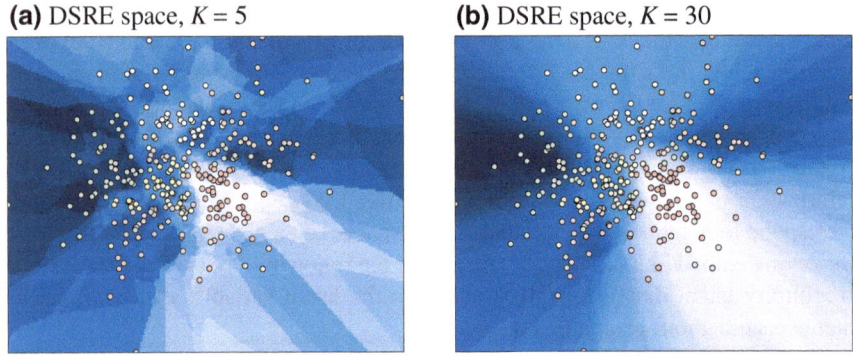

Fig. 8.2 Visualization of DSRE space $e(\cdot, \mathbf{y}_1, \mathbf{X})$ w.r.t. the first pattern $\mathbf{y}_1$ for $K = 5$, and $K = 30$ after a run of UNN with $N = 300$ embedded patterns. **a** DSRE space, $K = 5$. **b** DSRE space, $K = 30$

The PSEA optimization approach is based on the following two ideas:

1. Iteratively construct a solution (an embedding $\mathbf{X}$) to cope with the large number of free parameters, and
2. perform PSO-like black box search steps in each iteration to embed the latent point at an optimal position.

As the problem to minimize $E(\mathbf{X})$ scales linearly with the number of patterns N, which may be a very large number in practice, the iterative solution construction is the key concept for efficiently learning the manifold. The approach is described in the following (cf. Algorithm 1).

Algorithm 1 Particle Swarm Embedding Algorithm

1: **input: Y**, K, κ
2: **repeat**
3:     choose $\mathbf{y} \in \mathbf{Y}$
4:     look for closest pattern $\mathbf{y}^*$ with latent position $\mathbf{x}^*$
5:     **for** $i = 1$ **to** κ **do**
6:         update velocity (cf. Eq. 8.4)
7:         update latent position (cf. Eq. 8.3)
8:         evaluate $E(\mathbf{X})$ or $e(\mathbf{x}', \mathbf{y}, \mathbf{X})$
9:         update best position $\tilde{\mathbf{x}}$
10:    **end for**
11:    embed $\tilde{\mathbf{x}}$
12:    $\mathbf{Y} = \mathbf{Y} \backslash \mathbf{y}$
13: **until Y** $= \emptyset$

In each step the pattern that has to be embedded is randomly chosen $\mathbf{y} \in \mathbf{Y}$. In the particle swarm step we seek for the optimal position, where the particle $\mathbf{x}$ should be embedded. For this reason, a loop of PSO-like steps is repeated for κ iterations:

$$\mathbf{x}' = \mathbf{x} + \mathbf{v}' \tag{8.3}$$

with velocity

$$\mathbf{v}' = \mathbf{v} + c_1 r_1 (\tilde{\mathbf{x}} - \mathbf{x}) + c_2 r_2 (\mathbf{x}^* - \mathbf{x}) \tag{8.4}$$

Here, $\tilde{\mathbf{x}}$ is the best position w.r.t. the DSRE the latent particle has found so far, and $\mathbf{x}^*$ is the latent position of the embedded pattern $\mathbf{y}^* \in \hat{\mathbf{Y}}$ that is closest to the pattern $\mathbf{y}$ that we want to embed:

$$\mathbf{x}^* = \arg \min_{i=1,\dots,|\hat{\mathbf{Y}}|} \|\mathbf{y} - \mathbf{y}_i\|^2, \tag{8.5}$$

with the Euclidean distance $\|\cdot\|^2$. The parameters $c_1, c_2 \in [0, 1]$ are constants that define the orientation to the best latent particle, and the closest already embedded one. Variables $r_1, r_2 \in [0, 1]$ are uniform random values. Figure 8.3 illustrates the particle swarm embedding step. The new candidate latent point $\mathbf{x}'$ is generated with velocity $\mathbf{v}'$, and the two scaled vectors.

In the following, we analyze the PSEA variant that takes into account the reconstruction error $e(\cdot, \mathbf{y}, \mathbf{X})$ (cf. Eq. 8.2) of the pattern $\mathbf{y}$ that has to be embedded. A greedy, but slower variant of PSEA is possible that employs the overall DSRE (cf. Eq. 8.1) for each latent position.

Fig. 8.3 Illustration of PSEA: The new candidate latent point $\mathbf{x}'$ is generated with velocity $\mathbf{v}'$, and the two scaled vectors $\tilde{\mathbf{x}} - \mathbf{x}$ and $\mathbf{x}^* - \mathbf{x}$

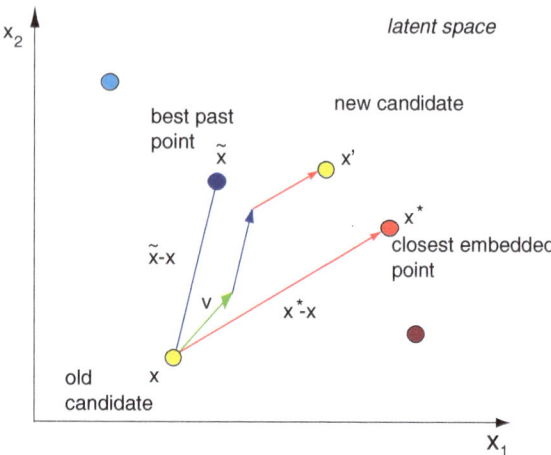

8.4 Experimental Analysis

We analyze the results of the novel PSEA experimentally. To evaluate the quality of the embeddings we employ the DSRE and a co-ranking matrix measure introduced by Lee and Verleysen [16]. It is based on the comparison of ranks (sorting w.r.t. distances from patterns) in data space and latent space. It defines a co-ranking matrix $\mathbf{Q}$ that explicitly states the deviations of ranks in data and latent space, c.f. [16] for a definition of $\mathbf{Q}$. In this matrix rank errors correspond to off-diagonal entries. A point $\mathbf{y}_j$ with lower rank w.r.t. a point $\mathbf{y}_i$ in latent space is called *intrusion*, a higher rank is called *extrusion*. From the co-ranking matrix the following quality measure can be derived that counts the number of proper ranks within a neighborhood of size K:

$$E_{NX}(K) = \frac{1}{KN} \sum_{k=1}^{K} \sum_{l=1}^{K} q_{kl} \tag{8.6}$$

This term restricts the measure to neighborhoods of size K. High values for E_{NX} show that the high-dimensional neighborhood relations are preserved in latent space, a perfect embedding achieves a value of one.

First, we analyze the influence of neighborhood size K on the results of PSEA, LLE and ISOMAP on two test data sets, i.e., *Digits* and *Boston*. For the PSEA, we choose the following settings. The particle swarm embedding process runs $\kappa = 50$ iterations. The initial velocity is randomly generated with a Gaussian distribution $\mathbf{v}_0 = \mathcal{N}(0, 1)$, the initial position starts from the latent position of the closest embedded point $\mathbf{x}_0 = \tilde{\mathbf{x}}$. The constants are both set to $c_1 = c_2 = 0.5$. Table 8.1 shows the experimental results w.r.t. the DSRE and E_{NX} for the settings $K = 5, 10, 15,$ and 30. Each PSEA experiment has been repeated 25 times. The best results, i.e., low DSRE and high E_{NX} are shown in bold, the second best are shown in italic

Table 8.1 Comparison of DSRE and E_{NX} with PSEA (mean values of 25 runs with standard deviation), LLE, and ISOMAP on the two test data sets *Digits* and *Boston*

	PSEA		LLE		ISOMAP	
K	DSRE	E_{NX}	DSRE	E_{NX}	DSRE	E_{NX}
Digits						
5	**15.87 ± 0.23**	**0.47 ± 0.01**	24.17	0.25	16.67	0.41
10	**18.77 ± 0.29**	**0.42 ± 0.01**	19.29	0.41	18.96	**0.42**
15	20.89 ± 0.64	0.40 ± 0.01	*19.98*	*0.44*	**19.52**	**0.47**
30	*24.17 ± 0.48*	*0.39 ± 0.01*	25.511	0.34	**21.97**	**0.51**
Boston						
5	**29.81 ± 1.86**	**0.45 ± 0.01**	45.29	0.30	*34.06*	*0.42*
10	**37.35 ± 6.40**	**0.43 ± 0.03**	*62.81*	0.29	81.57	*0.35*
15	*53.59 ± 2.94*	*0.40 ± 0.03*	69.35	0.20	**44.24**	**0.43**
30	53.03 ± 3.23	041 ± 0.04	*33.32*	*0.55*	**27.69**	**0.66**

numbers. The results show that a low DSRE correlates with a high E_{NX}. The DSRE is increasing with the neighborhood size. PSEA achieves the best results of all methods in case of small neighborhood sizes $K = 5$, and $K = 10$ on both data sets. In case of larger neighborhoods, ISOMAP shows better results, but PSEA still computes competitive embeddings, and achieves the second best results in half of the cases. LLE and ISOMAP win in performance for larger neighborhoods. The results of LLE are worse than the results of PSEA in three of the four cases, in particular E_{NX} tends to be much worse. Surprising is the bad result of ISOMAP on the *Boston* data set for $K = 10$.

In Fig. 8.4, we compare the PSEA results of embedding $N = 500$ patterns of the *Digits* data set employing varying settings for neighborhood size K and invested numbers of iterations κ. The figures show that for a small neighborhood size of $K = 1$ the embedding are worse that for the larger setting $K = 30$. Also if more search is invested, i.e., $\kappa = 100$ iterations instead of $\kappa = 20$, the quality of the embeddings cannot be improved significantly. For the larger neighborhood size $K = 30$, the lower two figures show that reasonable embeddings have been computed. More search invested ($\kappa = 100$) even slightly deteriorates the embedding with outliers. This is probably due to the effect that too greedy embeddings in each step can produce locally optimal outliers that deteriorate the overall scaling.

8.5 Conclusions

In unsupervised regression, the optimization problem of placing latent variables scales with the number of patterns and becomes impractical for large data sets. In this chapter, we have introduced a novel optimization approach that is based on the hybridization of iteratively constructing a solution and PSO-like optimization in each iteration. The proposed method belongs to the first particle swarm approach that

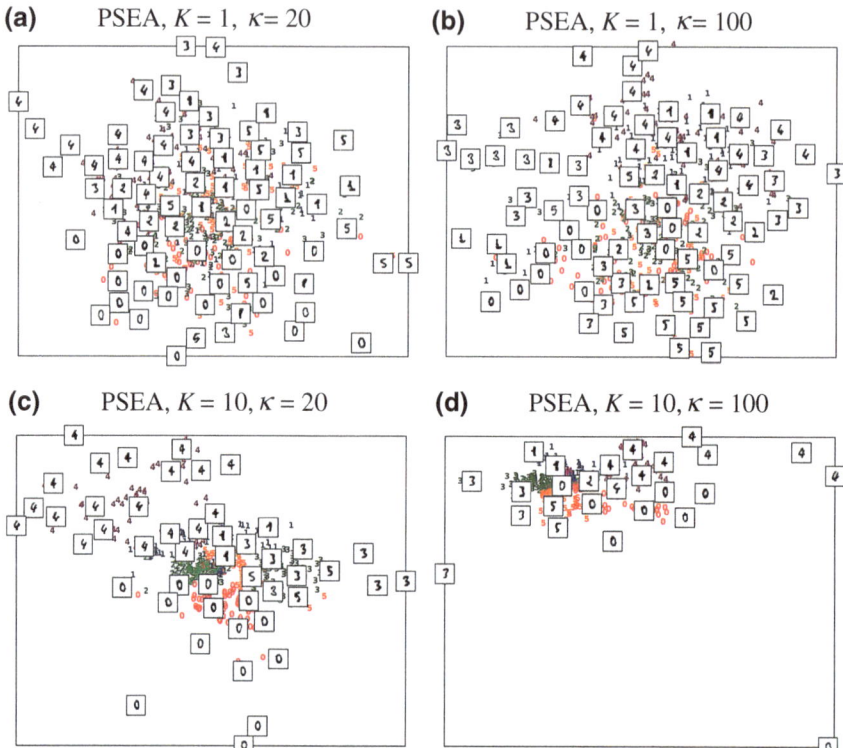

Fig. 8.4 Comparison of embeddings of 500 patterns, and 6 classes of the *Digits* data set. PSEA results for **a** $K = 1, \kappa = 20$, and **b** $K = 1, \kappa = 100$. **c** $K = 30, \kappa = 20$, and **d** $K = 1, \kappa = 100$

allows learning of low-dimensional embeddings from high-dimensional patterns. The results are competitive to embeddings of established methods like LLE and ISOMAP. The experiments have shown that the PSEA embedding fulfills conditions like neighborhood preservation and low DSRE. As extension of the PSEA approach, it is reasonable to parallelize the embedding process and thus allow to learn embeddings of large data sets. Another prospective research direction is to employ further DR criteria for the fitness evaluation of the optimization process like kernel density regression criteria.

References

1. J. Kennedy, R. Eberhart, Particle swarm optimization, in*Proceedings of IEEE International Conference on, Neural Networks* pp. 1942–1948, 1995
2. Y. Shi, R. Eberhart, A modified particle swarm optimizer, in *Proceedings of the International Conference on, Evolutionary Computation* pp. 69–73, 1998

3. S.T. Roweis, L.K. Saul, Nonlinear dimensionality reduction by locally linear embedding. Science **290**, 2323–2326 (2000)
4. J.B. Tenenbaum, V.D. Silva, J.C. Langford, A global geometric framework for nonlinear dimensionality reduction. Science **290**, 2319–2323 (2000)
5. P. Meinicke. Unsupervised Learning in a Generalized Regression Framework. PhD thesis, University of Bielefeld, 2000
6. P. Meinicke, S. Klanke, R. Memisevic, H. Ritter, Principal surfaces from unsupervised kernel regression. IEEE Trans. Pattern Anal. Maching Intell. **27**(9), 1379–1391 (2005)
7. N.D. Lawrence, Probabilistic non-linear principal component analysis with gaussian process latent variable models. J. Mach. Learn. Res. **6**, 1783–1816 (2005)
8. S. Tan, M. Mavrovouniotis, Reducing data dimensionality through optimizing neural network inputs. AIChE J. **41**(6), 1471–1479 (1995)
9. O. Kramer, On unsupervised nearest-neighbor regression and robust loss functions, in *International Conference on Artificial, Artificial Intelligence* pp. 164–170, 2012
10. Y. Kao, K. Cheng. An ACO-based clustering algorithm. *In Ant Colony Optimization and Swarm Intelligence (ANTS)* (Belgium, Springer, 2006), pp. 340–347
11. A. Abraham, C. Grosan, V. Ramos,(eds.), *Swarm Intelligence in Data Mining, Volume 34 of Studies in Computational Intelligence* (Springer, New York, 2006)
12. M. O'Neill, A. Brabazon, Self-organizing swarm (SOSwarm) for financial credit-risk, Assessment, 2008
13. T. Kohonen, *Self-Organizing Maps* (Springer, Berlin, 2001)
14. L. Herrmann, A. Ultsch. The architecture of ant-based clustering to improve topographic mapping. *In Ant Colony Optimization and Swarm Intelligence (ANTS)* (Belgium, Springer, 2008) pp. 379–386
15. F. Gieseke, K.L. Polsterer, A. Thom, P. Zinn, D. Bomanns, R.-J. Dettmar, O. Kramer, J. Vahrenhold, Detecting quasars in large-scale astronomical surveys, in *International Conference on Machine Learning and Applications (ICMLA)* pp. 352–357, 2010
16. J.A. Lee, M. Verleysen, Quality assessment of dimensionality reduction: Rank-based criteria. Neurocomputing **72**(7–9), 1431–1443 (2009)

Appendix A
Test Problems

This work makes use of experimental test problems to evaluate the introduced approaches. Here, we give a survey of the used test suites that range from optimization problems to machine learning test data sets.

A.1 Optimization

The optimization problems introduced in the following are used for the experimental analysis of the evolution strategies and the Powell ES. The experimental analysis concentrates on typical test problems known in literature on optimization.

OneMax

Maximize

$$\text{OneMax}(\mathbf{x}) = \sum_{i=1}^{N} x_i \quad \text{with} \quad \mathbf{x} \in \{0, 1\}^N. \tag{A.1}$$

The optimal solution is $\mathbf{x} = (1, \ldots, 1)^T$.

Sphere Model

Minimize

$$f_{\text{Sp}}(\mathbf{x}) = \sum_{i=1}^{N} x_i^2 \quad \text{with} \quad \mathbf{x} \in \mathbb{R}^N, \tag{A.2}$$

in matrix form

$$f_{\text{Sp}}(\mathbf{x}) = \mathbf{x}^T \mathbf{x} \tag{A.3}$$

with properties:

O. Kramer, *A Brief Introduction to Continuous Evolutionary Optimization*, SpringerBriefs in Computational Intelligence, DOI: 10.1007/978-3-319-03422-5, © The Author(s) 2014

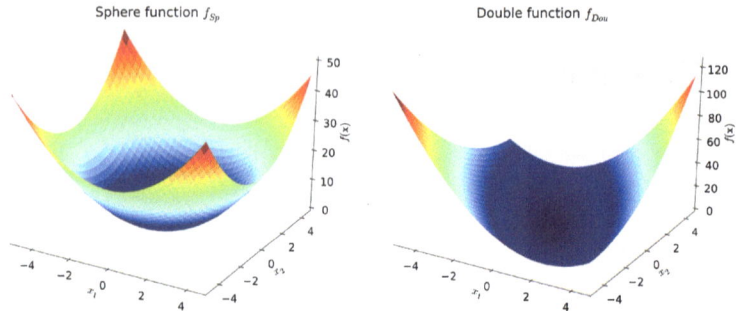

Fig. A.1 *Left* plot of the unimodal *Sphere* function. *Right* plot of the multimodal *Doublesum* function

- unimodal, separable,
- scalable,
- minimum $\mathbf{x}^* = (0, \ldots, 0)^T$ with $f(\mathbf{x}^*) = 0$.

Doublesum

Minimize

$$f_{\text{Dou}}(\mathbf{x}) = \sum_{i=1}^{N} \left(\sum_{j=1}^{i} (x_j) \right)^2 \quad \text{with} \quad \mathbf{x} \in \mathbb{R}^N \tag{A.4}$$

with properties

- unimodal, non-separable,
- scalable,
- minimum $\mathbf{x}^* = (0, \ldots, 0)^T$ with $f(\mathbf{x}^*) = 0$ (Fig. A.1).

Rosenbrock

Minimize

$$f_{\text{Ros}}(\mathbf{x}) = \sum_{i=1}^{N-1} \left(100(x_i^2 - x_{i+1})^2 + (x_i - 1)^2 \right) \quad \text{with} \quad \mathbf{x} \in \mathbb{R}^N \tag{A.5}$$

with properties:

- multimodal for $N > 4$,
- non-separable, scalable,
- very narrow valley from local optimum to global optimum,
- minimum $\mathbf{x}^* = (1, \ldots, 1)^T$ with $f(\mathbf{x}^*) = 0$. For higher dimensions, the function has a local optimum at $x = (-1, \ldots, 1)^T$.

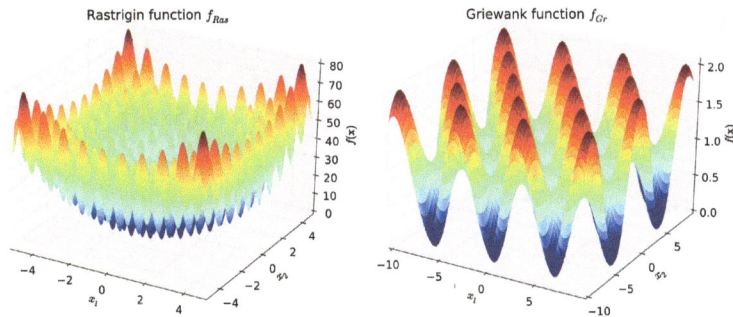

Fig. A.2 *Left* plot of the *Rastrigin* function. *Right* plot of the multimodal *Griewank* function

Rastrigin

Minimize

$$f_{\mathrm{Ras}}(\mathbf{x}) = \sum_{i=1}^{N} \left(x_i^2 - 10 \cos(2\pi x_i) + 10 \right) \quad \text{with} \quad \mathbf{x} \in \mathbb{R}^N \tag{A.6}$$

with properties

- multimodal, separable,
- large number of local optima,
- scalable,
- solution space $\mathbf{x} \in [-5, 5]^N$,
- minimum $\mathbf{x}^* = (0, \ldots, 0)^T$ with $f(\mathbf{x}^*) = 0$.

Griewank

Minimize

$$f_{\mathrm{Gri}}(\mathbf{x}) = \sum_{i=1}^{N} \frac{x_i^2}{4,000} - \prod_{i=1}^{N} \cos\left(\frac{x_i}{\sqrt{i}}\right) + 1 \quad \text{with} \quad \mathbf{x} \in \mathbb{R}^N \tag{A.7}$$

with properties

- multimodal, non-separable,
- scalable,
- minimum $\mathbf{x}^* = (0, \ldots, 0)^T$ with $f(\mathbf{x}^*) = 0$ (Fig. A.2).

Kursawe

Minimize

$$f_{\text{Kur}}(\mathbf{x}) = \sum_{i=1}^{N} \left(|x_i|^{0.8} + 5 \cdot \sin(x_i)^3 + 3.5828 \right) \tag{A.8}$$

with properties

- highly multimodal

TR-Tangent Problem

Minimize

$$f_{\text{TR}}(\mathbf{x}) = \sum_{i=1}^{N} x_i^2 \quad \text{(N-dim. \textit{Sphere} model)} \tag{A.9}$$

constraints

$$g(\mathbf{x}) = \sum_{i=1}^{N} x_i - t > 0, \quad t \in \mathbb{R} \quad \text{(tangent)} \tag{A.10}$$

- for $N = k$ and $t = k$,
- minimum $\mathbf{x}^* = (1, \dots, 1)^T$ with $f(\mathbf{x}^*) = k$.

Problem $f_{2.40}$

Problem $f_{2.40}$ minimizes $f_{2.40}(\mathbf{x}) = -\sum_{i=1}^{5} x_i$ subject to the following six constraints

$$g_k(\mathbf{x}) = \begin{cases} x_k \geq 0, & \text{for } k = 1, \dots, 5 \\ -\sum_{i=1}^{5} (9+i)x_i + 50{,}000 \geq 0, & \text{for } k = 6. \end{cases} \tag{A.11}$$

minimum $x^* = (5{,}000, 0, 0, 0, 0)^T$ with $f(x^*) = -5{,}000$.

Problem ZDT1

Problem ZDT1 minimizes the two objective functions $f_1(\mathbf{x})$ and $f_2(\mathbf{x})$ with

$$f_1(\mathbf{x}) = x_1, \tag{A.12}$$

and

$$f_2(\mathbf{x}, \mathbf{z}) = g(\mathbf{z})h(f_1(\mathbf{x}), g(\mathbf{z})), \tag{A.13}$$

with

$$g(\mathbf{z}) = 1 + \sum_{i=1}^{N} z_i/N, \tag{A.14}$$

and

$$h(f_1(\mathbf{x}), g(\mathbf{z})) = 1 - \sqrt{f_1(\mathbf{x})/g(\mathbf{z})} \tag{A.15}$$

Problem ZDT2

Problem ZDT2 is like ZDT1, except

$$h(f_1(\mathbf{x}), g(\mathbf{z})) = 1 - (f_1(\mathbf{x})/g(\mathbf{z}))^2 \tag{A.16}$$

Problem ZDT6

Problem ZDT6 is a bi-objective problem with

$$f_1(\mathbf{x}) = 1 - e^{-4x_1} \sin^6(6\pi x_1) \tag{A.17}$$

with

$$g(\mathbf{z}) = 1 + 9 \left(\sum_{i=1}^{N} z_i/N \right)^{0.25}, \tag{A.18}$$

$$h(f_1(\mathbf{x}), g(\mathbf{z})) = 1 - (f_1(\mathbf{x})/g(\mathbf{z}))^2 \tag{A.19}$$

and

$$f_2(\mathbf{x}, \mathbf{z}) = g(\mathbf{z})h(f_1(\mathbf{x}), g(\mathbf{z})) \tag{A.20}$$

For the definition of problems ZDT3 and ZDT4, we refer to Coello et al. [1].

A.2 Machine Learning Problems

Boston

The Boston housing data set stems from 506 census tracts of Boston in 1970. It consists of $N = 506$ patterns with $d = 13$ features (positive real values), e.g., *proportion of owner-occupied units built prior to 1940* and *weighted distances to five Boston employment centers*. The original data has been published by Harrison and Rubinfeld [2]. The data set was taken from the *StatLib* library, which is maintained at Carnegie Mellon University.

Fig. A.3 Visualization of a
collection of images from the
Digits data set

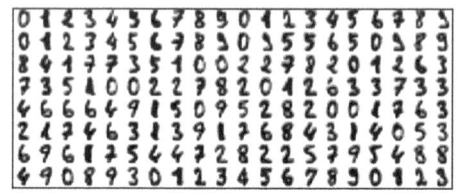

Digits

The *Digits* data set [3] comprises handwritten digits and is often employed as reference problem related to the recognition of handwritten characters and digits. Figure A.3 shows a collection of images from the *Digits* data set.

References

1. C.A.C. Coello, G.B. Lamont, D.A. van Veldhuizen, *Evolutionary Algorithms for Solving Multi-Objective Problems. Genetic and Evolutionary Computation Series*, 2nd edn. (Springer Science+Business Media, New York, 2007)
2. D. Harrison, D. Rubinfeld, Hedonic prices and the demand for clean air. J. Environ. Econ. Manage. **5**, 81–102 (1978)
3. J. Hull, A database for handwritten text recognition research. IEEE Trans. Pattern Anal. Mach. Intell. **5**(16), 550–554 (1994)

Index

O. Kramer, *A Brief Introduction to Continuous Evolutionary Optimization*,
SpringerBriefs in Computational Intelligence,
DOI: 10.1007/978-3-319-03422-5, © The Author(s) 2014